GUIDE TO
BIRDS
OF BRITAIN AND EUROPE

HÅKAN DELIN AND LARS SVENSSON

and illustrate
HÅKAN DELIN, MARTIN ELLIOTT, PETE
LARS SVENSSON and DA

Håkan Delin is a medical doctor, living north of Stockholm. He is generally regarded as one of the best field ornithologists in Sweden and is also a highly accomplished bird artist, who painted the owl plates in *The Birds of the Western Palearctic* (Oxford).

Lars Svensson is a Swedish ornithologist. He is author of *Identification Guide to European Passerines* and *Collins Bird Guide*, several papers on bird identification and editor of many books on birds. He is a member of the taxonomic sub-committee of the British Ornithologists' Union and the Swedish Taxonomic Committee.

PUBLISHER *John Gaisford*
COMMISSIONING EDITOR *Steve Luck*
EXECUTIVE ART EDITOR *Mike Brown*
DESIGNER *Chris Bell*
PRODUCTION *Sally Banner*

First published in 2007 by Philip's,
a division of Octopus Publishing Group Ltd,
2–4 Heron Quays, London E14 4JP

Based on a field guide originally created
by Bertel Bruun and Arthur Singer

Copyright © 2007 Philip's

Copyright © 2007 Håkan Delin,
Lars Svensson

Copyright © 2007 Paul and Alan Singer,
Peter Hayman

ISBN-13 978-0-540-08969-7
ISBN-10 0-540-08969-9

A CIP catalogue record for this book is available
from the British Library.

Printed in China

FRONT COVER : *tl* Grey Partridge and Red-legged
Partridge; *tc* Aquatic Warbler; *bl* Little Stint;
bc Tawny Owl; *r* Fulmar
BACK COVER : Grey Heron distribution map;
Bewick's Swan

Other titles from Philip's

Philip's publish a range of natural history
titles in the same format as this book,
including:

*Philip's Guide to Butterflies of Britain and
  Ireland*
Philip's Guide to Deserts
Philip's Guide to Fossils
*Philip's Guide to Freshwater Fish of Britain
and Northern Europe*
Philip's Guide to Gems
Philip's Guide to Minerals, Rocks and Fossils
Philip's Guide to Mountains
*Philip's Guide to Mushrooms and Toadstools
  of Britain and Northern Europe*
Philip's Guide to Oceans
Philip's Guide to Seashells of the World
*Philip's Guide to Seashores and Shallow Seas
  of Britain and Ireland*
Philip's Guide to Stars and Planets
Philip's Guide to Trees of Britain and Europe
Philip's Guide to Weather
Philip's Guide to Wetlands

For details of Philip's products
website: **www.philips-maps.co.uk**
email: **philips@philips-maps.co.uk**

PREFACE

When the predecessor to this book, 'The Hamlyn Guide to Birds of Britain and Europe', first appeared in 1970 it set a new standard for field guides to birds in Europe. Based on the highly successful American counterpart published in 1966, it was Bertel Bruun (text) and Arthur Singer (illustrations) who provided European birdwatchers with a modern, easy-to-use tool. Its concept with all vital information for each species – text, plate and map – appearing on the same spread, has since been followed by other guides. 'The Hamlyn Guide to Birds…' was revised several times, and in 1986 we were invited to rewrite the text completely and became the authors of the book. We also helped improve some of the illustrations and asked the talented Swedish artist Dan Zetterström to redo all the plates for waders, skuas, gulls and terns, groups which we felt required more accurate artwork in order to be helpful. His achievement meant a clear rise in the standard of the book.

The well-known British publisher of maps and textbooks, Philip's, has now acquired the rights to this title and asked us to update it and suggest improvements. The task has not been entirely easy, and we confess to having hesitated initially. Much has changed in the more than 35 years since the book was originally designed, and some things would be too costly to alter now. We have seen a minor revolution in avian taxonomy, with many new species recognised, and skills in the art of bird identification have been improved dramatically. Information on previously little known birds is now readily available on the Internet. Digital photography through telescopes ('digiscoping') has enabled amateur birdwatchers to contribute in a field where previously only specialised photographers with expensive equipments operated. By this development we now have access to pictures of virtually all birds in our part of the world. All these advances in ornithology cannot be reflected in a new version of the Hamlyn guide. But we have tried to take into account the most important novelties, concentrating on the needs of everyday birdwatchers rather than specialists.

First of all we have focused on Europe. All breeding species in Europe are given full coverage, as are those which migrate through or are vagrants from Asia and North America and which are fairly regularly found. Extreme rarities and species which are exclusive for Asia Minor and North Africa have been omitted to allow more space for describing the common European species, and to be able to treat the newly split ones. Split species now described under separate headings are e.g. Green-winged Teal, Balearic and Yelkouan Shearwaters, Yellow-legged and Caspian Gulls, several warblers, Southern Grey Shrike and Corsican Finch.

Advances and changes in the taxonomy of birds also affect the order in which the species are arranged. This fieldguide is one of the first to embrace the new insight that among European birds, wildfowl and gamebirds are the most basic on the evolutionary tree, and hence they are placed first in the book. More recently evolved orders and families are placed after in a so-called natural sequence.

No less than 43 plates (32%) have been newly commissioned, or have been composed by existing artwork from other sources. Martin Elliott has painted new plates for several seabirds, diving ducks, large gulls and some other images inserted on old plates. Håkan Delin has painted new plates for Anser geese, divers, grebes, bustards, cranes, auks, owls and woodpeckers, and a number of other images. Peter Hayman has allowed us to use his paintings for 'The Complete Guide to the Birdlife of Britain & Europe', a Mitchell Beazley imprint, to enable improvements of several plates, mainly of passerines and in particular for larks, pipits, accentors, warblers and some tits, and provided odd images for a variety of other groups.

The text has been revised to reflect the new taxonomy following several species splits in recent years, but also to mirror advances in identification techniques. Still, we remind the reader that in order to keep this guide truly portable, many finer details have been left out. (Anyone who wants to take his bird identification to a higher level is referred to more comprehensive books or specialist identification papers in birding journals.) The restrictions are not necessarily seen as a drawback. What is offered in this book is sufficient for the overwhelmingly most common birdwatching situations; it can almost be a relief to have a limited and realistic number of species to choose from when encountering a suspected Sedge Warbler in your home patch or Redstart in your garden.

We thank Jane and Ben Carpenter, who kindly read through the revised text and offered sensible advice. We should add that David A. Christie translated the Swedish text upon which 'The Hamlyn Guide to Birds of Britain and Europe' was based. We learnt much from this and hope that some of it is still evident in the text of this new book.

All the maps have been revised for this edition. Note that through their small size, any finer details cannot be accurately shown. The maps are meant to convey a general appreciation of the European occurrence at large of each bird.

Håkan Delin, Lars Svensson

CONTENTS

The coverage of this guide is the main part of Europe, the only exceptions being the easternmost parts of Russia and the northern slopes of the Caucasus. In the east, the border chosen is the 45th meridian, running from the Kanin peninsula in the north to the foothills of the central Caucasus in the south. From there, the border runs west along the northern foothills of the Caucasus to the Black Sea, through the Bosporus and Aegean Sea including all Greek islands (e.g.

Lesbos and Crete), further including Malta and all other European islands in the Mediterranean Sea. The border runs through the straits of Gibraltar and northwards, including also Iceland and Svalbard (Spitzbergen). The birdlife of Turkey east of the Bosporus, of Cyprus, the Levant, North Africa, Canary Islands, Madeira and Greenland is not treated (despite the fact that some of these areas are visible on the distribution maps).

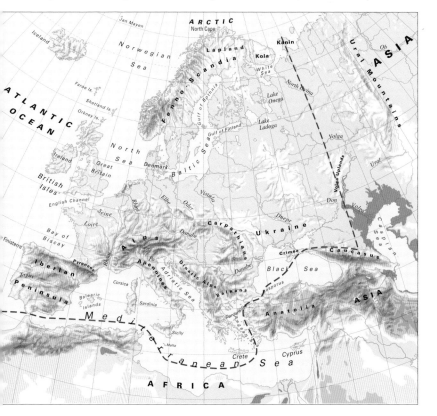

INTRODUCTION

Anyone who becomes interested in birds is to be congratulated – a rich and enjoyable pastime is guaranteed for the rest of his or her life.

Birds can be studied in many ways: through daring and difficult exploits, or quietly and calmly; while travelling to remote and exotic countries, or merely in areas close to home; alone and pioneering, or in a friendly and sociable group; systematically and scientifically, or just for fun.

Regardless of how you choose to pursue your study of birds, it is highly recommended that you know what you are actually looking at; you need to identify the birds, and often to ascertain their sex and age as well. We hope that this book will be a useful tool in this endeavour.

What this book contains – and what is not included

This book treats well over 500 species of birds. It follows therefore that the artwork and the text can cover only the normal and reasonably commonly seen; rare exceptions and most other 'oddities' have had to be omitted. Still, as the reader can see, the artwork is very comprehensive: every species is depicted in the various plumages which are normally met with, and flight pictures have been added where appropriate.

Eclipse plumage

Ducks moult their flight-feathers – all quills simultaneously - in the summer and then become flightless. During this vulnerable period the males replace their brightly coloured mating plumage with a discrete so-called eclipse plumage, which offers good camouflage. Since the eclipse plumage is worn mostly during a period when the birds lead a more secluded life, and also is very similar to the female plumage, it has not been depicted.

One way of telling eclipse males from the similar females is to look at the bill: the male retains his bill colour (the Mallard drake yellowish, the Red-crested Pochard drake red, etc.). The wing-pattern, too, often gives away the males: a male Steller's Eider in eclipse, which has moulted its body-feathers to brown-black, still has its white patches on the upperwing; it is then (before shedding the remiges) surprisingly similar to a Black Guillemot in flight.

▼ *Steller's Eider, male in eclipse plumage – at long range resembling a Black Guillemot.*

Juvenile plumages

In a number of passerines, the young have a first plumage, the juvenile plumage, which is already partly or completely moulted by the end of the summer. In many species it is basically similar to that of the adult female, in others so different that it is conspicuous and may even cause identification problems in the field. Most of these different-looking plumages have been included on the plates, but owing to lack of space not quite all. One example is the juvenile plumage of the Dunnock.

Thanks to the comprehensiveness of the plates, we have been able to avoid burdening the species accounts with complete plumage descriptions. Instead we have focused on critical characters, which have been put in italics to make them stand out. Also, more space has been available for voice descriptions.

Songs and calls

Being familiar with the songs and the calls of the birds is important for every birdwatcher. An early-morning walk through the forest in late spring can be a completely frustrating experience for the beginner: the bird songs fuse to become an anonymous chorus, and the members of the choir slip away in the canopy. Conversely, a spring night by a marsh or a lake, when birds are migrating or establishing their territories, can be a thrilling event for the knowledgeable listener. The experienced ornithologist indisputably notices and identifies just as many birds by ear as by eye.

In effect, the voices of the birds – songs, calls and alarm notes – are often quite decisive for the identification of a number of species: most warblers and pipits, the two treecreepers and many waders, to mention some obvious groups. And for nightjars, many owls, crakes and other birds with nocturnal and secretive habits, the voice is almost the only means available for identification.

A short-cut to learning bird voices is to listen to recordings or to benefit from the company of a knowledgeable friend. Often, such help, when most needed, is unavailable; you are left to steal and stalk and compare what you hear with what you see. This is a pleasant pastime but requires both alertness and patience. And if you are let down, the written voice descriptions in the bird-guides might help.

In this book, we have concentrated on those songs and calls which are most commonly heard at normal range (thus not only at closest range) and which help identification. This means that a lot of minor calls have been omitted, not only because of lack of space but for instructive purposes as well. For instance, we want to stress the rule that the flying Red-throated Diver is vocal but the flying Black-throated silent; therefore we deliberately say nothing of the fact that the latter species occasionally gives calls in flight (though still not resembling those of the Red-throated).

With very few exceptions we have chosen to render each call in only one way, again for instructive reasons, even though variations certainly occur.

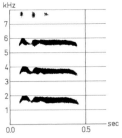

kHz

▲ *Sonogram of the begging call of a young Long-eared Owl – a drawn-out, mournful 'piii-eh', audible over 1 km.*

Sonograms or just letters?

In some modern ornithological handbooks, graphs of the vocal utterances, known as sonograms (sound-spectrograms), are used. Without denying the value of such, we have chosen the more conventional method of general descriptions in words and phonetic renderings with letters. We believe that sonograms require much experience and practice before they can be correctly interpreted; and they are space-consuming. These are two good reasons for sonograms not being used in this book.

There is also the risk that sonograms convey a false impression of precision, appearing like fingerprints of unique proof-value. But the fine shades and exact kilohertz values of the sonograms cannot of course replace the good judgement of the interpreter. For instance, the variation in calls within a species is rarely apparent when two closely related species are compared with the aid of sonograms.

Without sonograms in this book, we still think that our readers are pretty well equipped. We have devoted more space to voice descriptions than can be found in most other comparable bird books. We have attempted to use metaphorical comparisons and descriptive adjectives which are easy to understand and which are meant to make the phonetic renderings in letters (inevitably rather personal) more vivid. And for the renderings, an effort has been made to use vowels, consonants and punctuation in a consistent and meaningful way.

Dialects

A complication with voice descriptions in a book covering the whole of Europe is that birds, too, have dialects – the Blackbirds of Greece do not chatter just like those in England. By and large we have refrained from going into this aspect. Only in those cases where the dialects are not even similar, or where resemblance to other species occurs, have we made exceptions. The songs of the Redwing and Ortolan Bunting and the rain call of the Chaffinch are a few examples.

▼ *Black Stork in strong light – note how pale it can appear.*

Optical illusions

The plates, complemented by the descriptions in the species accounts, should provide a fair general picture of what the birds look like. But birds do not always look like they should. Light, visibility, background and, in the case of flying birds, wind conditions play an important part in our perception of them. A few general remarks are called for.

Effects of light

With a strong back-light, necks appear narrow – the light 'eats away' the contours: a perched

Golden Eagle can look almost turkey-like. Viewed against the sun, glossy black plumages can gleam whitish too: a soaring Black Stork does not always look so black above! Heavily overcast weather with poor light subdues colours – brown becomes blackish, etc. Overcast but lighter skies will instead enhance colours and patterns in a surprising way: alighting Bean Geese and soaring Lesser Spotted Eagles, for instance, can appear rather strongly patterned in certain lights. Strong light combined with haze makes birds appear pale: a Hobby may appear

▲ *Contrasts are important for size evaluation. Birds which stand out most from the background appear largest. Feral Pigeons in this simplified example.*

pale grey above like a Barbary Falcon. Strong sunlight creates deep shadows which can be mistaken for black colour: note, for instance, how difficult it is to see the rufous underwing colour of a Collared Pratincole in flight in sunshine. The faint light of dawn makes birds look surprisingly small and quick-winged: an Eider can look like a Common Scoter. And in fog all birds appear large.

Contrasts
The great importance of the background can be confirmed when looking at two Feral Pigeons, one white and the other black, flying side by side: the white one definitely appears the larger of the two when the background is dark, the black one suddenly larger when the background is switched to light. The Feral Pigeons are easily recognised regardless of this size illusion. But when we see a pale winter-plumaged Red-throated Diver in flight in strong light against a dark sea, the contrast may create a size illusion which will make us – with beating hearts – consider one of the larger and rarer diver species.

▼ *Shelduck against a dark background. The resulting image is confusing.*

Birds with a pied pattern can have their true shape distorted by the background. A flying Shelduck can, for instance, appear very strange if seen against a dark forest: neckless, short-winged and long-tailed. And a White-billed Diver on a pale sea appears to have lost its bill!

Evaluating size
Every identification of a bird in the field initially involves a size determination of some kind, from broad limitations such as 'between thrush and dove' to sophisticated attempts such as 'slightly larger than a Lesser Whitethroat'. As a guide, the length of each species (from tip of bill to tip of tail, bird flat) is given in the accounts. For certain species, the wingspan (wingtip to

▲ *A White-billed Diver will appear bill-less on a light-coloured sea.*

wingtip, wings naturally extended) has been added.

The experienced birdwatcher will not wish to discard size as an aid to identification, but is nevertheless aware of its many limitations and pitfalls; some of these have been mentioned already under 'Optical illusions', and others are treated below.

Prejudices about size

Size evaluation can be affected by preconceived but false conceptions. Although we know that the Raven and the Great Black-backed Gull are large birds, many of us still have a tendency to underestimate their true size since their shapes are so close to those of their smaller and more familiar relatives; we refuse to realize how much larger than a crow or a Common Gull they are. Accordingly we find that a Buzzard appears small when it is chased by a Raven, and the Black-throated Diver seems to be of modest size when a Great Black-back alights nearby.

Normal variation in size

The size variation within a species should be mentioned, too. The considerable size difference between Ruff and Reeve is well known. So, too, is the difference between males and females of Black Grouse and Capercaillie (males largest) and of Peregrine, Merlin, Goshawk and Sparrowhawk (females largest) – the sexes often appear like two different species.

Size differences among cranes, geese, cormorants and gulls are perhaps not quite so well known. And young of geese and cranes (just as those of bustards and gamebirds) are still not fully grown in the autumn. If a formation of migrating Barnacle Geese in the autumn contains a single Anser goose which is clearly smaller in size, this should not automatically lead to cries of 'suspect Lesser White-fronted

▶ *A pursuing Raven makes a Buzzard look surprisingly small.*

Goose'; it could be a small young female White-fronted.

There is a normal size variation of roughly 6–12% within each sex and age-group of all species. This is often overlooked, even by experienced birdwatchers, and can cause misidentifications. The smallest possible Little Stint among other Little Stints of normal size is bound to attract much attention from birdwatchers, craving for rarities, before its true identity is established.

▲ *We expect a Black-throated Diver to be a large bird: a large male Great Black-backed Gull nearby confuses our preconception.*

Wingbeat speed and its effects on size impression
Strong winds force flying birds to make quicker movements, and they may then appear much smaller than they really are: an Osprey can look like a gull, a Goshawk like a Sparrowhawk, etc.

Several birds, such as divers, birds of prey, waders, terns, auks, larks and finches, have display flights with slow-motion wingbeats. Since we associate a bird's wingbeat speed with a certain size of bird, we frequently overestimate the size of birds which perform such display flights and can be misled, for a moment. A displaying Calandra Lark in the distance can appear as large as a small raptor. Also, active moult of flight-feathers can alter not only the normal speed of wingbeats, but also the silhouette, both affecting the size impression, and hence deceive the human observer.

Hybrids and other oddities
The birdwatcher's knowledge and imagination are really put to the test by hybrids, miscoloured birds, and birds in unusual transitional plumages.

Hybridization is fairly frequent among wildfowl (especially among those in parks). A male hybrid with two dabbling-duck species (genus Anas) as parents will generally be strongly patterned and cause little confusion. Hybrids of the smaller diving ducks (Aythya) on the other hand can be much more tricky, since they often resemble related American species in a confusing way.

Hybrids are presumably occurring fairly frequently among smaller birds, too, but many of the few which are noticed are probably discarded as aberrant individuals of one or other of the parent species involved.

Miscoloured birds sometimes resemble related species.

▲ *Hybrid Lesser White-fronted goose × Barnacle Goose.*

13

▲ *Partially albinistic Robin which can be confused with a female Red-flanked Bluetail.*

▲ *Young Starling in transitional stage of moult to first-winter plumage; only the head remains of the juvenile plumage.*

Partially albinistic Starlings, Blackbirds and Robins can, for instance, be confused with Rose-coloured Starling, Ring Ouzel and female Red-flanked Bluetail, respectively.

Birds moulting from one plumage to a completely different one can look strange at certain stages. The juvenile Starling which is moulting to its first winter-plumage has, for a period, a plain brown head while the body is black with large white spots. This is well known and causes no problems. But anyone who encounters a one-year-old Hen Harrier male starting to moult from its worn juvenile plumage to its first adult-type plumage may have some trouble: such a plumage appears pale sandy-grey at a distance and is not well described in the literature. The identification must then to a large extent rest on shape, size, movements and general impression, the so-called jizz (an expression of obscure origin, but commonly used by bird-watchers), something which cannot be learnt from books but has to be practised in the field. As so often, practice makes perfect, and to recognise the unusual you have to be fully familiar with the normal and common.

Birdwatching strategy

The early bird catches . . .

For most of us, a trip to watch birds away from home involves some planning, and the spending of some money and time, and naturally we want as much pleasure as possible from it. One way is to be out early. Each time we experience the hectic activity among the birds at dawn and sunrise, we regret all the similar moments we have missed. Remaining until dusk can to some extent compensate for a missed dawn, for many birds are very active in the evening as well.

The importance of the weather

It is always a good idea to let the weather govern the planning of field trips, especially in the migration seasons. Warm spring weather and cold spells in the autumn trigger the birds' departure, and they prefer to migrate in fair weather with good visibility and tail-winds. Nevertheless, more birds are sometimes seen in rain and storms. Fair days with favourable winds result in migration high up and out of sight (exceptions are larger birds of prey, and migrants such as cranes and geese which fly in formation). Bad weather or strong head-winds force many birds to halt, and large roosting flocks can be seen at favourable spots. We must also keep in mind that it is the weather conditions in the area of departure that are the essential factor, and this can be a long distance away.

Seabirds are naturally forced closer to land in strong onshore gales, and owls are more vocal on calm, mild nights following cold weather and rain.

The best results are therefore produced by different weather conditions in different habitats.

Detect more with the help of other birds

Birds are difficult to spot. This is true especially for birds of prey, which glide rapidly through the terrain or pass high overhead. Helpful indications of their presence come at times from other birds, so often more sharp-sighted than us. 'Lesser' bird-of-prey species are revealed by high-pitched, drawn-out notes from tits, the agitated twittering of Pied Wagtails, the 'glit, glit' calls of Swallows or the hard 'kyett' calls of Starlings. Larger birds of prey are often brought to our attention by the grating 'krrrr' calls or furious 'kraa' calls of Carion/Hooded Crows (the first expressing 'annoyed superiority', the other 'furious fright'), the 'krra krra krra' of Ravens, the persistent '**klee**-u' of Common Gulls or the deep '**glaa**-o' of Herring Gulls.

▲ A Dotterel on the ground tilts its head and attentively watches the sky – often a sign of a passing bird of prey.

A chorus of alarm calls is also what usually reveals an owl on its roost in daylight. Loudest alarm is often given by the thrushes.

It is safe to assume that en masse take-off by flocks of birds on open fields signals an approaching bird of prey. A more subtle hint is when perched or ground-feeding birds scan the sky with head tilted: usually there is a raptor to be found high among the clouds (or at least a gull or a Raven).

Optical equipment

Those who are seriously interested in birds will want to acquire a good pair of binoculars as soon as possible. And if the interest is strong enough to stay and become an important part of one's life, then it is natural to add a telescope to the equipment; it should be used on a steady tripod with a fluid-head-type joint, allowing smooth scanning. Tripods made of carbon fiber are more expensive but also both lighter to carry and more steady in the wind.

A telescope cannot be recommended strongly enough; every kilogram carried pays off! Not only will you increase your identification range considerably, but you will also get a much closer view into the everyday lives of the birds and detect interesting details in their behaviour. And beautiful scenes are offered without disturbing the birds: a few Swallows on a dry branch, a brood of ducklings on a river bank, a Buzzard at the edge of the forest lit by the morning sun, for example.

▼ Hunting Buzzard on its favourite perch.

Avoid disturbance

Every responsible birdwatcher makes it a point of honour to avoid disturbing the birds. It is sometimes said that 'bird-lovers are the birds' worst enemies'. Fortunately this is not true. The ornithologist is observant, and can interpret the calls and the behaviour of the birds and realise in time when they need to be left in peace. Unsuspecting sunbathers, anglers and others are surely a much greater threat.

Birds are most sensitive to disturbance when breeding, especially in the early stages. Certain species will desert the nest permanently if chased off newly laid eggs. Well-beaten tracks to nests can help winged or mammalian nest-predators to find the eggs. The eggs will get cold and the embryos die if the parents are kept away from their nest for lengthy periods in chilly weather. And strong sun can kill small young. The young of, for example, waders and gamebirds leave the nest shortly after hatching and crouch in the grass when a human being approaches, relying on their cryptic plumage; they run the risk of being trampled to death.

But birds can suffer from disturbance even outside the breeding season. Geese, for instance, need to graze for most of the brief winter day to survive, and just as important, to build up surplus fat for days with unsuitable weather, and for development of the eggs. And some birds of prey visit their nest all year and might, if disturbed, choose a less suitable site when the spring comes.

Protect the rare

Theft of young wild birds of prey for use in falconry is a sad and outrageous activity, as is the perverted collecting of eggs of rare birds, and killing of e.g. owls and raptors for

▲ Be observant in the breeding season and avoid harmful disturbance. A parent bird with food in its bill, anxiously calling all the time, means that you should move away a little.

mounting. We immediately contact the police if we suspect anybody of planning such criminal activities. But we can also protect the nests of rare species by keeping quiet about their whereabouts. And we should also bear in mind that attractive birds are not immune just because they breed far away in foreign countries, out of our own reach: if we pass on details of a Gyrfalcon's site in Iceland, it will receive several visitors, year after year; the remoteness of Iceland will not discourage.

Not only must the interests of the birds come first, but we must also respect those of landowners and local residents. Common sense and considerate behaviour will usually suffice. If an extreme rarity occurs at a site, followed by the inevitable invasion of 'twitchers' (keen birders keeping a numerical life-list of bird species seen), it is desirable that an able leader steps forward to direct the traffic, negotiate with authorities and arrange some compensation for the landowner, and keep up the goodwill of birdwatchers in general.

Identification is not all

Finally, let us repeat: the principal aim of this book is to facilitate identification of species, sex and age. Most people interested in natural history would, however, agree that this is but a first step. A bird is certainly more than an appearance and a few calls. An exhausted Long-eared Owl in a bush by the sea, newly arrived, is a memorable sight, and one which can be studied in depth; but the image of the Long-eared Owl can also be complemented with the emotional atmosphere of a misty early spring night when the pair hoots its melancholy duet. Or the memory of warm summer evenings when the owls hunt for voles along flowering ditchbanks. And a single White-fronted Goose on the ground, however confidently identified and aged it may be, cannot evoke the dream of the tundra in the way a large 'V' of White-fronts making headway high up while calling in chorus can.

Birds do not come alive and catch our imagination until we learn something about their habits and behaviour. Only then can we speculate about their fascinating adaptations and prospects in life. For this endeavor this book serves as a tool.

▼ *Long-eared Owl, just arrived over the sea, resting in first available tree.*

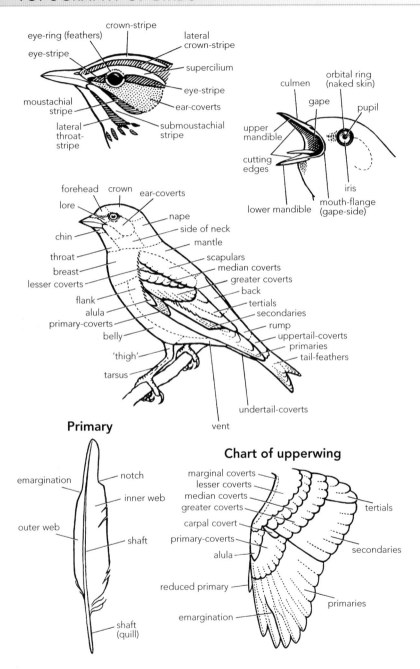

eye-ring (feathers)
crown-stripe
lateral crown-stripe
eye-stripe
supercilium
eye-stripe
moustachial stripe
ear-coverts
lateral throat-stripe
submoustachial stripe

orbital ring (naked skin)
culmen
gape
pupil
upper mandible
cutting edges
iris
lower mandible
mouth-flange (gape-side)

forehead   crown   ear-coverts
lore
nape
side of neck
chin
mantle
throat
scapulars
breast
median coverts
lesser coverts
greater coverts
flank
back
alula
tertials
primary-coverts
secondaries
belly
rump
'thigh'
uppertail-coverts
tarsus
primaries
tail-feathers
undertail-coverts
vent

Primary

emargination
notch
inner web
outer web
shaft
shaft (quill)

Chart of upperwing

marginal coverts
lesser coverts
median coverts
greater coverts
tertials
carpal covert
primary-coverts
alula
secondaries
reduced primary
emargination
primaries

Distribution Maps

Abbreviations and symbols

The status in the British Isles		Miscellaneous	
R	resident	♂ imm.	immature male
S	summer visitor	♀ imm.	immature female
W	winter visitor	L	total length (in cm)
P	passage visitor	W	wingspan (in cm)
V	vagrant	juv.	juvenile, young

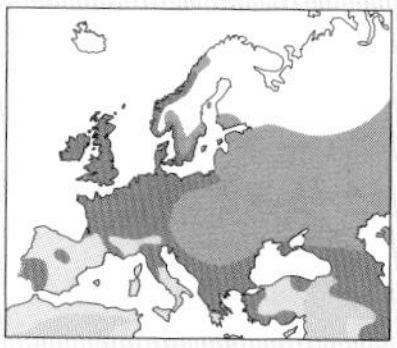

Breeding areas

Established migration areas

Wintering areas

Occurs all year round

For some species, the irregular winter and summer limits are shown as broken blue or red lines.

How the book is arranged

Brief general descriptions are given for each order, family, subfamily or other main group. After this introduction follow the species accounts.

The order of the various groups reflects their evolutionary history, starting with those which are thought to be oldest and ending with the most recently evolved. A novelty in this book is that the wildfowl is placed first followed by the gamebirds. Recent research using comparisons of the genomes has shown that these two groups are older than the others and hence should come first among the European birds. Another surprising new insight is that grebes and flamingoes are each others closest relatives, and so it is logical to place these two together. Using genetics in taxonomy is still relatively new, and we can expect more rearrangements in the future.

Only in a very few instances have unrelated species deliberately been placed together, the reason being that it is practical to be able to compare them. Thus, the Small Button-quail is placed next to the unrelated Quail and the Snow Finch appears with the Snow Bunting, both with striking white wing pattern.

As to the species descriptions, text, map and illustrations for each bird are kept together on the same spread. The text gives English and scientific name, a rough indication of size (L = total length in centimetres from tip of bill to tip of tail when the bird is stretched out, W = wingspan with wings fully but naturally stretched), abundance, habitat, field-marks (the most useful of which are put in *italics*), habits and vocalisations. For species recorded in Britain a letter or combination of letters at the end of the account give a quick hint at its status.

Rendering calls with letters is difficult and rather personal, but we have still attempted to convey with words how they sound and how they differ from similar calls by using comparisons and adjectives. When the German letter 'ü' (as in 'Lübeck') has been used it is meant to signify a vowel with a pitch intermediate between 'u' and 'ee', for which there is often a need in order to achieve accurate renderings. A call 'trrüee' is upwards-inflected whereas 'feeü' is down-slurred.

WILDFOWL (*order Anseriformes, family Anatidae*)
Wildfowl are, together with gallinaceous birds (grouse and pheasants), now regarded as the most basal group on the evolutionary tree among European birds, and this group is hence placed first. Wildfowl are aquatic and have webs between the three forward-pointing toes. They have long necks and relatively narrow, pointed wings as well as short legs. Their often slightly flattened bodies are well insulated with down beneath the feathers. Most build simple nests lined with down on the ground in thick vegetation. The newly-hatched young are covered in down, and can both walk and swim a few hours after hatching.

THE SWANS are the largest of the wildfowl, have long necks – as long as their bodies. All three European species are white. The young are brownish-grey. All have to run along the surface in order to get airborne. Herbivorous. Clutches 3–7 eggs. **p.22**

GEESE are intermediate between swans and ducks in size and appearance and in other respects, but form a distinct group. Sexes alike. The geese are heavier and have longer necks than the ducks. Most species have legs placed farther forward than the ducks and the swans, an adaptation to their habit of grazing on land. They are good fliers, making long migration flights between stops, forming striking flight patterns (Vs, arcs and lines). Congregate at traditional moulting sites. Clutches 3–8 eggs. **p.24**

SURFACE-FEEDING DUCKS are appreciably smaller than swans and geese, have flatter bills and shorter legs. Known also as dabbling ducks, they are found on ponds, lakes, reservoirs and slow-flowing rivers, where they feed on vegetable matter. Strong fliers, they take off steeply from the water. Often have distinct and diagnostic wing patterns. Sexes dissimilar. Males brilliantly coloured, females in most species camouflaged in brown. In summer males acquire female-like plumage ('eclipse'), while the flight feathers are moulted and

Whooper Swan

Greylag Goose

Mallard – surface feeding duck

the ability to fly is lost. Largest in the group are the shelducks, and in these the plumage differences between the sexes are small. They have longer necks and look more goose-like than do other dabbling ducks. Hybrids between species not uncommon. Exotic extralimital species may occur but are in most cases escapes from zoos and wildfowl collections. Clutches 6–12 eggs. **p.30**

DIVING DUCKS all have a lobe on the hind toe. They are excellent divers and have the legs placed far back. Bills flattened. In most cases they run along the surface of the water to take off. The smaller diving ducks nest mostly on inland waters in N Europe and winter principally on reservoirs, lakes and coastal waters in W Europe. They live mainly on aquatic vegetation and molluscs. Clutches of 5–12 eggs. The heavy diving ducks are often found further out to sea than the smaller diving ducks and live to a large extent on molluscs. They nest both along sea coasts and on inland lakes. In winter they are seen almost exclusively on sea coasts. Hybrids between different species of diving duck occur rarely and give rise to difficult problems of identification. Clutches 4–8 eggs. **p.38**

THE SAWBILLS have bills which are well adapted to holding slippery fish. Their bills are narrower than in other wildfowl and equipped with a hooked tip on the upper mandible and sawtooth-like lamellae along the edge. The sawbills take off with long runs like the heavy diving ducks. Nest in hollow trees (with the exception of Red-breasted Merganser). Clutches 5–10 eggs. **p.48**

STIFFTAILS are small, plump ducks that live on lakes and swamps. They are characterised by their quite long, stiff tails. Clutches 5–10 eggs. **p.48**

examples of diving ducks

Pochard

Eider

Red-breasted Merganser – typical sawbill

White-headed Duck – stifftail

Swans (subfamily Cygninae)

Very large, heavy, white long-necked birds. Sexes alike. Ungainly, waddling walk. Majestic on the water. Up-end like surface-feeding ducks in order to browse on the bottom. Build large nest of vegetable matter. Clutches of 3–5 (8).

Mute Swan

Mute Swan *Cygnus olor* L 150, W 210. Numerous and widespread in Europe. Nests in reeds on lowland lakes, gravel-pits, sluggish rivers and canals (even in loose colonies), often close to human presence. Non-breeders gather in large flocks. In winter in flocks on the coasts. Fierce territorial combats in which dominant males drive off intruders with wing-splashing rushes and 'slides' along the water. Can behave rather aggressively, even towards man, more particularly so during the breeding season. Heavy, weighs on average 8–12kg. When swimming it holds the *neck in a graceful S-shape* with the bill pointed downwards, often also with the wings raised in shape of a shield. The tail is long and sharply pointed. Adult's bill *orange-red with black knob*, juvenile's greyish-mauve with dark at the base. The juvenile is more variegated brown and white on the wings than juvenile Whooper and Bewick's Swans. Comparatively *silent* but far from 'mute'. The adults give an explosive lashing 'rhepp' and a more subdued 'heeorr'. Hisses when using threat behaviour. *A loud singing buzz with each wingbeat* is heard from adult Mute Swan *in flight*. **R**

Whooper Swan

Whooper Swan *Cygnus cygnus* L 155, W 215. Nests in northernmost Europe in swamps and tundra and taiga lakes. In recent decades has begun to nest, uncommonly and locally, further and further south. Habitually very shy at breeding site, but the recent new colonists in the south considerably more fearless than those in the north. In winter usually along the coasts as well as on larger lakes and watercourses, may often graze on land. Swims with *upright neck* and never raises its wings like Mute Swan. This makes both it and Bewick's Swan quite easy to distinguish from Mute, even at longer range. When it up-ends the shorter, blunter tail is obvious. At close range the *yellow on the bill*, more extensive than in Bewick's Swan, can be seen. Juvenile distinguished from juvenile Mute Swan by more pale and cold grey colour, *paler bill* (largely grey-white) and by the silhouette, from juvenile Bewick's by size. Has *far-carrying call with melancholy tone, like blasts on a bugle*. The commonest flight call consists of three short blasts in rapid succession, 'klo-klo-klo'. When resting on the water, the Whooper often spins out the notes, sings. A chorus of large flocks is striking. *No singing noise from the wings* as in Mute Swan (only a slight and ordinary swishing). **W**

Bewick's Swan

Bewick's Swan *Cygnus columbianus* L 122, W 185. Least common of the swans, but locally some larger flocks. Nests on the arctic tundra. In winter and on migration frequents lakes, reservoirs, flooded grasslands and sometimes sheltered sea bays. A small version of the Whooper Swan but the *yellow on the bill is less extensive*, does not project forward in a wedge. *Neck proportionately shorter* than in Whooper Swan. Wingbeats slightly quicker than Whooper's. Often seen in family parties or large flocks. The juvenile has same plumage coloration as juvenile Whooper Swan, is more difficult than adult to identify by bill markings since the border between light and dark is indistinct. Calls are like Whooper Swan's but obviously higher-pitched. The singing chorus of drawn-out 'klah' notes at a distance sounds like clamouring Cranes. The loud cackling flight call is bent into a diphthong 'kleu', not straight like the Whooper Swan's. Neither does it have latter's tendency to three syllables but is monosyllabic (or disyllabic), thereby acquiring a yelping character, can recall geese. In quieter mood gives a muffled Whooper-like 'kokokoko'. **WP**

display

juv.

Mute Swan

Mute Swan
on nest

Whooper Swan

juv.

feeding

juv.

Bewick's Swan

Geese (subfamily Anserinae)

Geese are large birds with long necks. They feed on seeds, grass, aquatic plants and nutritious roots. Indefatigable fliers. Wingbeats composed. Fly in V-formation or in bent or diagonal line. Have curious habit of half-rolling and pitching when flying in to land. Sexes alike. Pair for life. Long-lived.

Canada Goose *Branta canadensis* L 90–100, W 165–180. N American species, split into several races of various sizes, the smallest nowadays treated together as a separate species, Cackling Goose *B. hutchinsii*. The large species introduced to Europe. Also spontaneous vagrants claimed in Britain. Breeds at inland waters, preferably large and open ones, also in forest districts. English population mainly sedentary, large Swedish one migrates to S Sweden and down to Netherlands. Feeds much in shallow water, like a swan, but also on pasture and arable fields. Not unlike an *Anser* goose at a distance, neck appearing dark, *pale breast striking*, but is larger and *longer-necked*; flight more majestic. Call a loud dissonant honk, 'rhot', at times in see-sawing duets; in flight disyllabic, second note in falsetto, 'rho-**üt**' (can also be rendered 'gah-**honk**'). **R**

Canada Goose

Brent Goose *Branta bernicla* L 60, W 115. Two races regularly seen in Europe. Dark-bellied race *bernicla* breeds in great numbers on Siberian coastal tundra, wintering on North Sea coasts. Light-bellied race *hrota* is sparse breeder on Svalbard, Greenland and NE Canada, wintering on Danish Jutland, in NE England and Ireland. Very rarely, birds of race *nigricans* ('Black Brant', by some regarded as separate species) from E Siberia and N Canada can be spotted in migrating and wintering flocks. Main winter food is eel-grass (*Zostera*) found on tidal mudflats. In shallow water up-ends like a duck. Rests on sea when tide is up. Sometimes grazes on coastal meadows, can then mix with Barnacles, otherwise flocks are unmixed. *Small, grey-black* goose with *gleaming white 'stern'*. Juveniles can acquire *white half-collar* in first autumn but retain white edges to wing-coverts into spring. In flight beats noticeably *narrower wings* at quicker pace than other geese, almost like Eider; goose impression caused mainly by length of wings. A few dark-bellied can be quite pale on upper flanks (beware not to take swimming such birds for light-bellied race). 'Black Brant' has *extensive white half-collar*, meeting at front of neck, and *contrasting white edge along blackish flanks*. Call a gargling 'r-rot'. **W**

Brent Goose

Barnacle Goose *Branta leucopsis* L 65, W 135. Breeds on foxproof cliff ledges and offshore islets, in three well-separated populations, in E Greenland, Svalbard and Novaya Zemlya (greatest numbers), wintering in W Scotland/Ireland, the Solway Firth and in the Netherlands, respectively. Began to nest on a few small islands off E Gotland (Sweden) in the 70s, now breeds in considerable numbers all over the Baltic. Grazes on coastal meadows in large flocks (usually unmixed). Juveniles similar to adults but lack crescentic barring on flanks typical of all adult geese. Migrating flocks often large, usually fly in Brent-like bow formation rather than in regular V, but with markedly *slower wingbeats than Brent*, equal to White-fronted Goose. In flight, at a distance, white head is difficult to discern, *breast-belly contrast* being a better character. Call a nasal, monosyllabic yelping 'gak', merging into a loud roar in large flocks. **W**

Barnacle Goose

Red-breasted Goose *Branta ruficollis* L 60, W 120. Breeds on Siberian tundra, in groups on steep river banks, preferably close to nest of Peregrine or Rough-legged Buzzard for protection. Winters mainly in SW Asia, but also in SE Europe, and odd birds join flocks of other geese and migrate to W Europe. Small, with *rather thick neck* and *very small bill*. *Broad white flank stripe on black body* is a better fieldmark at a distance (also in flight) than chestnut-red breast. Juvenile has a smaller chestnut cheek patch than adult. Call a shrill, staccato 'ki-kwi'. **V**

Red-breasted Goose

Canada Goose
(Only the larger species shown here)

swimming

juv.

Brent Goose

light-bellied race

light-bellied

dark-bellied

dark-bellied race

Barnacle Goose

Red-breasted Goose

Greylag Goose

Greylag Goose *Anser anser* L 75–85, W 147–170. The most widespread goose in Europe, nests chiefly in swamps and on reedy lakes but also e.g. on small islands on the sea coast, often in small colonies surrounded by loafing flocks of non-breeding immatures. Largest, *heaviest Anser* goose. Wings comparatively broad and blunt. Rather *large bill*. Pale pink legs. Greylag has, like Pink-footed Goose, *pale panels on forewings*, but Greylag's panels are paler, almost silvery-white, and contrast more sharply with secondaries and back. Also, *Greylag's wing is bicoloured below*: pale grey lesser and median coverts, and dark grey greater coverts and flight feathers (all other *Anser* geese have uniformly dark underwings). On the ground, Greylag never appears pale-backed like Pink-foot. On the other hand *head and neck are characteristically pale*. Individuals flying directly away from observer can be distinguished from other geese by grey lower back, which is in pale contrast to browner scapulars. Juveniles have dark nail to bill. Like other *Anser* geese is gregarious and readily mixes with other species. Pitches and freewheels in flight even more than Pink-foot when landing. Has, like other *Anser* species, a wide repertoire of calls, incl. really shrill ones, but the most typical is a nasal, cackling '**kyang**-ung-ung', the first syllable typically higher-pitched and more stressed. (Calls identical to those of domestic goose, of which Greylag is ancestor.) **RWP**

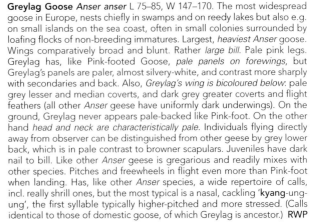

White-fronted Goose

Ageing of Anser geese

old

young

White-fronted Goose *Anser albifrons* L 60–73, W 130–160. Breeds on arctic tundra. A western population (race *flavirostris*; rather big, bill long and yellow-orange, plumage fairly dark) breeds in W Greenland and winters in NW Britain (numerous), an eastern one (race *albifrons*; slightly smaller than *flavirostris*, bill weaker and predominantly pink, plumage somewhat paler) breeds along the Russian arctic coast and winters from England (uncommon) diagonally across Europe to Turkey and eastwards. Adult White-front has *white blaze on forehead* and *black markings on belly*, can be confused only with Lesser White-front. Distinguished from latter by generally perceptibly larger size, larger bill, by fact that *forehead blaze is straight in side view and does not extend so far up on the forehead*, also by *lack of yellow orbital ring*. Juvenile distinguished from juvenile Lesser White-front mainly by lack of yellow orbital ring, slightly paler plumage and larger size. Since white forehead blaze is absent during first autumn and nail of bill is dark, juvenile can at distance be confused with Bean Goose, but is told from latter by *paler cheek with swarthy forehead and area around bill base* together with slightly narrower white feather edges on upperparts. Flight much as in other *Anser* geese. Does not appear pale-winged like Greylag and Pink-foot, but nevertheless has an obvious pale stripe formed by light-tipped upper wing-coverts, paler than in Bean Goose. Commonest call is characteristically high-voiced 'kyüyü', almost laughing; more conventional cackles also given. **W**

Lesser White-fronted Goose *Anser erythropus* L 56–66, 115–135. Very rare breeder in Fenno-Scandian mountains, in willows in upper birch zone. Once common now has dwindled to a mere 30 birds in N Norway. Has an eastern migration route to reach SE Europe and possibly Iran and Iraq. Young birds (colour-ringed) are reintroduced in Swedish Lapland, with Barnacles as foster parents, which then teach the young to migrate to the North Sea coasts. The young birds then return to Lapland on their own. Looks like a small version of White-fronted Goose, but has *small bill* and *steep forehead* (looks 'cute'), *white forehead blaze reaching far up* on fore-crown (often angled outline in side view), *yellow orbital ring* (visible also at distance) and *darker, more brown plumage* (though has same paler stripe on upper wing-coverts as White-fronted). Young birds lack the blaze and black belly-patches but has thin yellow orbital ring. Wingbeats fairly quick, a little quicker than other *Anser* geese, but nothing like the quick pace of equally large Brent Goose. The common flight call, 'kyiyi', is much more high-pitched (and more piping!) than that of White-fronted. **V**

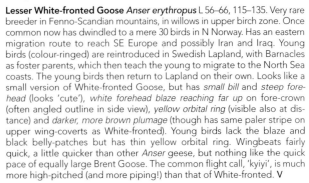

Lesser White-fronted

juv.

Greylag Goose

adult

race flavirostris

race albifrons

White-fronted Goose

juv.

adult

adult

juv.

**Lesser
White-fronted Goose**

adult

Pink-footed Goose

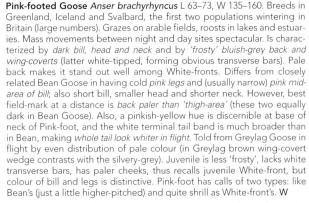

Pink-footed Goose *Anser brachyrhyncus* L 63–73, W 135–160. Breeds in Greenland, Iceland and Svalbard, the first two populations wintering in Britain (large numbers). Grazes on arable fields, roosts in lakes and estuaries. Mass movements between night and day sites spectacular. Is characterized by *dark bill, head and neck* and by *'frosty' bluish-grey back and wing-coverts* (latter white-tipped, forming obvious transverse bars). Pale back makes it stand out well among White-fronts. Differs from closely related Bean Goose in having cold *pink legs* and (usually narrow) *pink mid-area of bill*; also short bill, smaller head and shorter neck. However, best field-mark at a distance is *back paler than 'thigh-area'* (these two equally dark in Bean Goose). Also, a pinkish-yellow hue is discernible at base of neck of Pink-foot, and the white terminal tail band is much broader than in Bean, making *whole tail look whiter in flight.* Told from Greylag Goose in flight by even distribution of pale colour (in Greylag brown wing-covert wedge contrasts with the silvery-grey). Juvenile is less 'frosty', lacks white transverse bars, has paler cheeks, thus recalls juvenile White-front, but colour of bill and legs is distinctive. Pink-foot has calls of two types: like Bean's (just a little higher-pitched) and quite shrill as White-front's. **W**

Bean Goose

Bean Goose *Anser fabalis* L 68–80, W 142–165. Breeds on the tundra (race *rossicus*) and by taiga bogs (race *fabalis*) in N Europe. Winters on the Continent; rare in Britain. Closely related to Pink-footed Goose. Is, however, clearly darker and (especially *fabalis*) larger, longer-necked, larger-headed and longer-billed. Diagnostic character is Bean Goose's *orange legs and bill markings*. Tail has narrower white terminal band than in Pink-foot, looks generally darker from behind in flight. At longer range best told from Pink-foot by darker back (as dark as flank area) and wings – together with Lesser White-front has darkest wings of all *Anser* geese (though forewings with blue-grey tone and can appear fairly pale in slanting light). Note that many *fabalis* have white rim at bill base, though never as White-front. Can be confused with juvenile White-front, but *whole of head dark* (not only forehead and area around bill base), and colour of bill different. When grazing (dark head and neck hidden) can be difficult to pick out among Greylags, since both species are equally dark on back and almost the same size. Race *rossicus* is safely distinguished from *fabalis* only by shape of bill: *short*, still *stout* (bulkier than that of Pink-foot), with *bulging lower mandible* and *wide slot between halves* ('cigar bite'). Also, usually *much black markings* on bill (however, a criterion shared by 10% of *fabalis*). Is generally smaller than *fabalis* (though larger than White-fronted). Commonest call is a deep, nasal, disyllabic, jolting 'ung-unk', with more of a base tone than Pink-foot's. **W**

Bar-headed Goose *Anser indicus* L 75. Breeds in highlands of Central Asia, crosses Himalayas on migration to and from India. Escapes from zoos sometimes seen in the company of grey geese. Feral population in the Netherlands. Unmistakable: *white head with two black transverse bars on nape.* Body and wings very pale, look *off-white*, especially in flight. The juvenile is blackish on all of crown and nape.

Blue Goose
(variety of
Snow Goose)

Snow Goose *Anser caerulescens* L 65–80, W 135–165. In some cases a genuine visitor from North America, but most birds seen in W Europe are probably escapes from zoos and collections. Usually seeks the company of grey geese. Adult of the white variety (commonest) is easily recognised by *pure white plumage with black primaries.* Juveniles are sullied grey-brown but blackish primaries contrast well. The dark variety, the 'Blue Goose', has white head, blackish body and grey wing-coverts. Hybrids of unknown extraction may imitate this pattern but lack elongated drooping scapulars/greater coverts and have wrong colour/shape of bill. The call is a harsh monosyllabic 'keeh'. **V**

Pink-footed Goose

juv.

race rossicus

race fabalis

Bean Goose

juv.

race rossicus

race fabalis

Bar-headed Goose

juv.

juv.

Snow Goose

juv.

juv.

29

Surface-feeding ducks (family Anatinae)

Surface-feeding ducks, also called dabbling ducks, are often found in shallow water. They feed by 'up-ending' or by skimming the water surface with bill. Strong fliers, often rising almost vertically. Most can dive, but seldom do. The sexes are very different in plumage.

Mallard

Mallard *Anas platyrhynchos* L 56, W 95. Commonest and most widespread of the surface-feeding ducks, also the largest and heaviest. Identified in flight by *size, robust body, moderately quick wingbeats. Fine, high-pitched wing whistle characteristic of species.* Found in parks and on city canals, on rivers, ponds, lowland lakes, woodland swamps, upland waters etc. Very active at night, not least on migration in N Europe. Versatile in choice of nest site, readily accepting 'duck-baskets'. The males gather in flocks before the summer moult. In winter found in large flocks on the sea along northern coasts, elsewhere commonly inland. Males court females in winter when he gives a short weak whistle 'piu'. Female has a loud quacking call; drake a one-syllable, quieter, nasal and confident 'vehp'. **RWP**

Gadwall

Gadwall *Anas strepera* L 51, W 89. Rather uncommon but widespread in Europe. Breeds mostly on shallow open freshwater lakes and pools with reed cover or small overgrown islands. In winter also on reservoirs, gravel-pits, floodland – tends to avoid salt water. The male is comparatively *dull grey*, but on the water shows a characteristic black stern. In flight it is distinguished by a square-shaped *white, dark-framed speculum*. Female very like Mallard. When swimming her white belly is not visible, then look for *orange along the sides of comparatively thin bill*. In flight she has a characteristic white speculum, which is smaller and with a less prominent dark border than the male's. The male has a fairly low, slightly Corncrake-like 'rrep' call and a shrill 'pyee' in pursuit flight and courtship. Female's call is slightly higher-pitched than Mallard's. **RW**

Pintail

Pintail *Anas acuta* L male 71, female 56, W 89. Rather scarce as a breeding bird in W Europe, mainly on upland pools, lowland marshes and lagoons. In winter on sheltered estuaries, floodlands and nearby lakes. *Long-necked*, generally *slim build*, 'greyhound-like'. Characteristic *white rear edge to the brown speculum*. Male has grey and black bill, female grey. Male has a long pointed tail in breeding plumage. On spring migration seen mostly in pairs. Shy. Upstretched neck of alert male glistens among the aquatic vegetation. His spring call is a short whistle, 'kree', like Teal's but weaker. In autumn often joins flights of Wigeon (with which commonly associated in winter). **RWP**

Wigeon

Wigeon *Anas penelope* L 46, W 81. Not an uncommon breeder in north (taiga zone) on shallow, open fresh waters. On migration and in winter often in large flocks, mainly on coasts and on flooded grasslands. Grazes on grassy banks, on coasts, eats eel-grass. Migrating flocks form long lines. Fairly *long wings*, compact body, *pointed tail, short neck, slightly rounded head*. Younger males lack white wing panels, otherwise plumage like adults'. Female's wing markings rather insignificant, but because of the whitish innermost secondary (see fig. below left) can be confused with female Gadwall. All Wigeon plumages show *sharply offset snow-white belly*. Male has typical loud whistle, 'wheee-oo'. Night-migrating flocks give a yapping 'wip, wee-wee'. Female's call is a Goldeneye-like 'karr-karr-karr-…'. In autumn a snorting 'ra-karr' is heard. **RWP**

American Wigeon *Anas americana* L 51, W 87. Rare visitor to W Europe. Male typical with *white band on crown* and *green panel on side of head*, flanks and back rufous. Female very like female Wigeon but has paler and greyer head and more rosy tinge to flanks. *Underwing-coverts and axillaries white* (not brownish-grey as in Wigeon). The male has a whistle like Wigeon's but weaker and of three syllables, 'whee-whee-whew'. **V**

taking flight

landing

up-ending

dabbling

♂

♂

♀

♀

Mallard

♂

♀

♀

♂

Gadwall

♀

♂

♂

♀

Pintail

♂

♀

♂

♀

♂

Wigeon

♂

♀

♀

♂

♂

American Wigeon

31

Teal

Teal *Anas crecca* L 36, W 61. Fairly common and widespread, breeds on smaller usually fresh waters in uplands, lowland and coastal areas. In winter on shallow estuaries, saltmarshes, lakes and reservoirs. Mainly a nocturnal migrant. Gathers in large flocks. Drake very colourful in spring plumage but looks (like the female) generally dark at a distance, when characterised (apart from small size) mainly by *whitish-yellow patches on side of rump*. Female like female Garganey but has more evenly coloured side of head, a slightly *shorter bill*, usually with *a little yellowish-red at the base*, and *light patch on side of tail base*. See also wing pattern and under Garganey. Readily takes to the wing and manoeuvres to and fro above reeds and marshes in tight flocks with smooth flight like waders. It may then be further distinguished by *white wingbar in front of speculum*. The pale belly not very conspicuous. The male's call is a *far-carrying ringing whistle as clear as a bell*, 'kreek'. Female has a shrill croak, considerably more nasal than female Mallard's. **RWP**

Green-winged Teal *Anas (crecca) carolinensis* L 36, W61. The North American counter-part of our Teal is nowadays usually regarded as a separate species. It is a rare visitor to W Europe. The male has a *vertical*, not horizontal, *white stripe on the side of the body*. Also, weaker pale lines bordering green head-patch. The female is indistinguishable from the European Teal. Behaviour and call as Teal. **V**

Blue-winged Teal *Anas discors* L 38. Rare visitor from North America. The male has dark violet-grey head with a *gleaming white crescent* on each side, yellow-brown flanks dotted black, black rear behind a large white spot. Flying male at a distance appears rather all dark, only with strikingly *pale blue* (almost whitish) *wing panels*. Female resembles female Garganey but has less marked pattern on head (though has a whitish spot near base of bill), noticeably *longer bill*, yellowish feet and *much paler and bluer forewings* (almost as male), but lacks broad white rear line of speculum. **V**

Garganey

Garganey *Anas querquedula* L 38, W 63. Scarce on lowland lakes, mainly on small pools in marshy and flooded meadows. *Small* as Teal. In flight the male's wings appear whitish blue-grey at long range. Younger males have dark grey wings (pattern otherwise like older males). Female's forewings are brown-grey albeit not at all strikingly pale, do not differ much from those of female Teal. Best told from the latter in that the *rear white band bordering the speculum is obviously wider than the front one* (as in Pintail) instead of the other way round. On the water female distinguished from female Teal by a more patterned face: *light chin*, obvious also in side view (this is often the best mark!), *paler supercilium* and a light stripe below the dark eye-stripe (widening in front into a *light spot at bill base*), bordered below by *dark cheek markings*. Moreover bill is a little longer and all grey. Lacks pale patch at tail base, but has tertials edged a trifle paler. Also, usually swims with wingtips a bit lifted. The male's call is a *drawn-out dry crackling* with a hollow wooden ring: 'knerrek'; female's call is a Teal-like shrill and feeble croak. **SP**

Shoveler

Shoveler *Anas clypeata* L 51, W 79. Fairly common on very shallow lowland waters with surrounding cover. Feeds in flooded vegetation, often hidden out of sight. On water has 'front-heavy' appearance with the *long spoon-shaped bill* lowered. Also front-heavy in flight. The wings have large blue-grey panels (but not so pale and striking at long range as in male Garganey). The Shoveler is usually identified immediately in flight by the long bill. The female in flight shows an *all-dark belly* contrasting with white underwing-coverts. (Wigeon and Gadwall have white belly, while Mallard and Pintail are somewhere in between.) On rising, male's wings make a drumming noise. The male's spring call is a nasal double-note 'sluck-**uck**', heard mostly in the evenings. The female quacks at the same time, broadly with the same rhythm and emphasis: 'pe-**ett**'. **RWP**

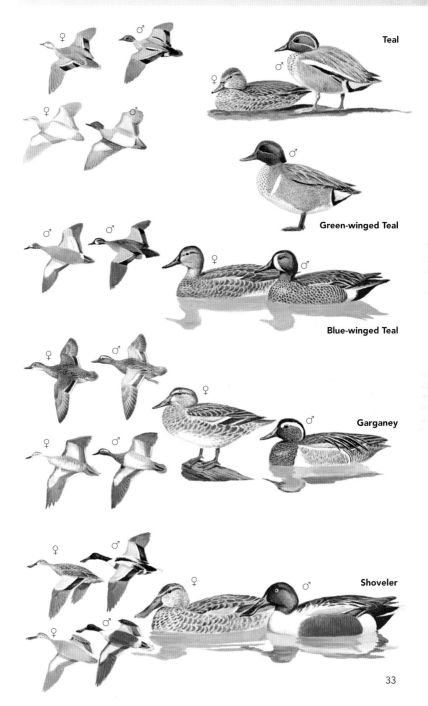

Teal

Green-winged Teal

Blue-winged Teal

Garganey

Shoveler

33

Marbled Duck

Marbled Duck *Marmaronetta angustirostris* L 40. Rare and local breeder in SW Europe. Prefers sheltered ponds and marshland with rich vegetation. Both sexes similar, light brown and with no very obvious distinctive features. Between Garganey and Wigeon in size. Best characteristics on swimming bird are striking *pale overall impression*, long pale tail, long neck and rather large round head with *dark area around the eye and dark*, quite long and *narrow bill*. In flight *long-winged* and can recall female Pintail. Lacks speculum, rear wing pale. Stays hidden and is difficult to approach. Seen mostly alone or a few together. The male's call is a quiet, high nasal 'jeeb', the female's similar, a double whistle 'pleep-pleep'.

Mandarin Duck

Mandarin Duck *Aix galericulata* L 46. East Asian species, introduced. Escaped birds have established feral populations in parts of England. Prefers ponds surrounded by trees, breeds in holes in trees. The beautiful male is unmistakable. The female is less strikingly coloured but note the characteristic head pattern; distinguished from the similar female Wood Duck *Aix sponsa* (in Europe found only in bird collections) by the absence of green on the rear crown and the different bill-outline at the base of the bill (see fig. below). In flight, *long tail* and *white belly* are striking. Usually feeds on land. **R**

Shelduck

Shelduck *Tadorna tadorna* L 60, W 110. Fairly common breeder in NW and SE Europe on flat, shallow coasts, locally also on inland lakes. Nests in burrows and under bushes near the shore. Male and female are similar in plumage, which is *white, reddish-brown and black with a green gloss*. Male has a large *red knob on the bill*, female a small pale red one. Immatures are considerably paler than adults, among other things lacking the rust-brown breast band, and have white chin and cheeks. Normally flies low over the water with arched wings moving in a rhythm half-way between duck and goose. During moulting they often form very large flocks, while at the same time the young on the breeding grounds which are still not able to fly also gather in flocks and are attended by only a few older birds. The male's spring call a high whizzing whistle: 'sliss-sliss-sliss-…'. The female's call is a characteristic straight whinnying 'gehehe-heheheh', and when nervous a nasal intense 'ah-ang', usually uttered in flight. Silent outside the breeding season. **RS**

Ruddy Shelduck

Ruddy Shelduck *Tadorna ferruginea* L 60. Breeds in SE Europe. Escapes from captivity also appear regularly outside the normal range, and have formed a large feral population in Switzerland. More terrestrial than Shelduck, breeding on steppe and in dry mountains. Fairly long-legged. Nests in hollows in the ground as well as in trees. Found in winter along rivers, by sandy lakeshores and on fields and steppes. The *orange-brown body* and the pale head are characteristic, *white panel on the wing striking in flight*. The male has narrow black neck band. Resembles Shelduck in build and behaviour. Flight also exactly like Shelduck's. Call a trumpeting 'galaw', somewhat like Canada Goose but more nasal and not so powerful (can recall plaintive cry of a donkey). Also has sonorous gurgling 'porrr'. **V**

Egyptian Goose

Egyptian Goose *Alopochen aegyptiaca* L 70. African species. Escaped birds have established rapidly growing populations in the North Sea countries. Note *large white wing panel*, both above and below, and the long legs. Sexes identical in plumage. Feeds mainly up on dry land, as geese habitually do. Harsh trumpeting calls. **R**

Mandarin Duck ♀

Wood Duck ♀

34

Marbled Duck

♀ ♂

Mandarin Duck

♀ ♂
♀
♂

♂ ♀ ♂ juv. **Shelduck**

♂

near nesting hole

♂ ♀ ♂ **Ruddy Shelduck**

♂

Egyptian Goose

Diving ducks (subfamily Aythyinae)

Diving ducks nest by lakeshores, on islands and in swampy areas. In winter they occur in flocks in sheltered bays, on larger lakes and in river mouths, but also further out to sea where several species feed on crustaceans and other small animals. They dive from the surface of the water, swim underwater, and run along the surface to take off.

Red-crested Pochard

Red-crested Pochard *Netta rufina* L 56. Scarce and local breeder in S and C Europe on brackish lagoons and reedy lakes. Escapes not uncommon. *Large*. Male has rather brilliant colours: *flanks are gleaming white*, head is yellowish-brown, crown feathers form an erectile crest, *bill is red* (even in female-like eclipse plumage). Female is superficially like female Common Scoter but paler, larger, has *pink band across outer bill*, like male has *very broad snow-white wingbar*. Behaviour resembles that of surface-feeding ducks, sits higher on water than scoters. **WP**

Scaup

Scaup *Aythya marila* L 46, W 79. Uncommon to scarce breeding bird on northern coasts (in Scandinavia on upland lakes in birch and willow zone). Most easily confused with Tufted Duck. The male however has *pale grey back*, which at a distance produces shining white 'amidships' appearance. Female has broad white band (sometimes tinged brown) around bill base and *in summer a pale mark towards the back of the cheek*, and is slightly paler than female Tufted, more reddish-brown (breast) and grey-washed (back). The Scaup is slightly larger, *at all times lacks crest on head and has less peaked forehead* (gives rounder head profile) and has slightly larger bill. Only nail of bill is black. **WP**

Ring-necked Duck *Aythya collaris* L 43. Very rare winter visitor to W Europe from North America. Male easily told from male Tufted by head shape and vertical white stripe on sides level with front edge of wing. Female has Pochard-like head pattern. Head shape and *grey* (not white) *wingbar* in both sexes is characteristic. **V**

Tufted Duck

Tufted Duck *Aythya fuligula* L 42, W 70. Common. Breeds on wide variety of lowland waters, incl. park lakes. Outside breeding season in large flocks on lakes, reservoirs, gravel-pits and sheltered coasts. Male characteristic, *black with white rectangle at the side and drooping crest*. Female has shorter crest and often has narrow white band at bill base (sometimes broad one: see Scaup). Belly brown during the nesting period. Often white under the tail (cf. Ferruginous Duck). Both sexes have yellow eye. Whole tip of bill black. Female's call is a repeated lively 'kerrb', male's spring call a giggling 'bheep-bhibhew'. **RW**

Pochard

Pochard *Aythya ferina* L 46, W 79. Breeds fairly commonly on well-reeded marshy lakes. In winter on lakes, reservoirs, gravelpits, sometimes sheltered estuaries. Male characteristic. Lacks whitish 'amidships' appearance of Scaup. Female more nondescript but has diffuse dark cheek patch between pale eye-stripe and pale chin. *Head shape triangular* with hefty bill and *flat forehead. Wings greyish-brown in flight*, palest in male. In breeding parties, males often outnumber females. Court female with hoarse 'bhee-bhee- . . .', also utter nasal, rising whistle (which may cut short with a 'chong'). Female's call is a harsh 'krrah, krrah,...'. **RSW**

Ferruginous Duck

Ferruginous Duck *Aythya nyroca* L 40, W 66. Breeds in S and C Europe on reedy lowland lakes. Rather unobtrusive. Slightly smaller than Tufted Duck, and *white wingbar is noticeably broader* and brighter in flight. Male is rich dark red-brown with *white eye* and gleaming *white undertail. White area of belly is smaller and completely enclosed by dark colour*. Female is dark greyish-brown with dark eye, has white undertail. Note *flat forehead and high crown*, and *rather long bill*, which with brilliant white wingbar are safest distinguishing marks compared with female Tufted. Female's call is a repeated, burring 'karri', with particular, high, almost ringing tone, quite different from female Tufted. **W**

Red-crested Pochard

♀   ♂

Scaup

♀   ♂

♀ *summer*   ♀ *winter 'Scaup-type'*   ♂

Tufted Duck

♀   ♂

Ring-necked Duck

♀   ♂

Pochard

♀   ♂

Ferruginous Duck

Diving ducks in flight

The diving ducks shown on the opposite plate have in common more or less prominent white or pale bands along trailing edge of the wing.

Red-crested Pochard *Netta rufina* Large and heavy, like a Mallard. The flight, too, recalls Mallard. But plumage pattern is invariably characteristic, there is rarely reason to hesitate about the species for either sex. Male unmistakeable. Even at longest range, the *very broad white wingbar* on upperwing is striking (underwing is entirely white). The *body is black below* with a peculiar pattern of *large, sharply marked white ovals on each side of the belly*. Note also the white fore-edge of the wing, and the presence of a large white patch on body at the fore wing join. The *coral red long bill* can be seen at reasonable ranges, as can the paler crown looking 'dyed blond'. The more discretely plumaged female is also easy to recognise in flight; she too has *very broad white wingbars*. That her *head is two-coloured* like on a female Common Scoter can also be seen at long range.

Scaup *Aythya marila* In flight recalls Tufted Duck, and practice and care is required for long-range identification, since the silvery back of the adult male or the white 'muzzle' of the female are then difficult to discern (a muzzle which many female Tufted Ducks display as well!). The Scaup has a *slightly more oblong head shape* (due to slightly longer head and bill) than the Tufted, and the *white wingbar is somewhat more prominent*, both characters requiring a reasonably close range. Long-range identification makes use of the particular 'jizz' of the Scaup: it is large and heavy enough to momentarily recall both Common Scoter and Wigeon – never so with Tufted! Often, the identification is greatly simplified by a few Tufteds (then strikingly small) being mixed in in a flock of Scaups. Flock formation, however, seems to offer little guidance between Scaup and Tufted. Note that both Scaup and Tufted fly higher in tail-winds than do Common Scoters.

Tufted Duck *Aythya fuligula* Small, only Teal being smaller among common species of ducks. *Wingbeats very fast*, almost like auks and grebes (whereas Teal beats its wings with hint of 'clipping' action). The *white wingbar is usually quite prominent*, even though it is fractionally less well-marked than in Scaup. That the *head is rather rounded* and proportionately large can usually be seen at long range. The tuft is on the other hand rarely visible in flight, not even when the birds are seen close. On migration, Tufted Ducks travel in small or moderately large flocks, in oblique lines or blunt 'Vs', and often at some height, in tail-winds frequently at considerable altitude. Flocks are more prone to veer in flight and make sudden shifts in altitude than relatives. Cf. also under Scaup.

Ring-necked Duck *Aythya collaris* Small like Tufted Duck. Note grey, not white, wingbars in both sexes. Head shape, which is so characteristic for swimming birds (slanting forehead and marked peak far back on crown), is hardly visible on flying birds.

Pochard *Aythya ferina* Medium-sized, clearly bigger than Tufted Duck and appearing *heavier and more long-stretched*. Wingbeats are fast, flight path usually quite straight, the Pochard gives you the impression of a 'cruising missile'. This can be enhanced by its habit to *hold its head rather low* (in common with Velvet Scoter), almost like a charging bull! On closer range the *sloping forehead* is evident. The wings are darker grey on their fore-part, paler at rear, but still hardly a species with obvious light wingbars as its congeners. The female is discretely patterned in brown and grey, the male obviously has more contrasts. The grey back of the male does not stick out as in the male Scaup (which has silvery back!), but blends more with the *uniformly dull grey of the upperwing*.

Ferruginous Duck *Aythya nyroca* Small, even slightly smaller than the Tufted Duck, but flight similar. Slightly more elegant shape with longer neck than Tufted. This rare duck is mostly seen on shorter flights above those small and reed-edged lakes where you already know that they breed. To see one on migration away from its breeding sites requires quite a bit of luck. Note the *extensively dark colour* in both sexes (that of the male an attractive deep chestnut-brown), the *broad white wingbars* (markedly broader and more shining white than in Tufted Duck), and the *sharply set off white belly patch*, with a *small white isolated patch also on the undertail*, separated from belly by a broad brown band. The *white eye of the male* is visible at longer range than the yellow eye of the Tufted male. However, the characteristic head shape with sloping forehead is generally of little help on flying birds.

Red-crested Pochard

♂

♀

Scaup

♂

♀

Tufted Duck

♂

♀

Ring-necked Duck

♂

♀

Ferruginous Duck

♂

♀

♂

♀

Pochard

Goldeneye

Goldeneye
imm. ♂ moulting

Goldeneye *Bucephala clangula* L 45, W 79. Fairly common breeder on waters in northern forests. Nests in Black Woodpecker holes and in nest-boxes. Hardy species, the last ones move south only to escape the ice. Winters along coasts, also on reservoirs and gravel-pits. Has tendency to disperse in smaller groups, does not pack together in large dense flocks like, e.g., Tufted Duck. Decidedly shyer than most other diving ducks. *Big head* with characteristic shape: *'triangular' with peak in centre of crown.* Male's *white loral patch* is visible at long range; flanks brilliant white, like drake Goosander but rear black. Both sexes have white wing panels like sawbills, but obviously *dark underwings*. Immature lacks the female's yellow eye and white neck ring, looks generally dark grey-brown but has white speculum (though smaller than in adult female) and the characteristic head shape of the species. Downy young are white-cheeked. For distinguishing marks in comparison with Barrow's Goldeneye, see latter. The male has a *musical whistling wing noise* – a well-known phenomenon of early spring in the northern countries is the characteristic paths of sound drawn across the night sky by migrating Goldeneyes. Migrants often rest on lowland lakes. When displaying, the male tosses his head backwards on to the back, splashes his feet and utters piercing rasps 'be-beeezh' (accompanied by a low, hollow, Garganey-like rattle). The female gives a grating 'berr, berr…' when she circles over the breeding territory. **WP**

Barrow's Goldeneye *Bucephala islandica* L 53. Very rarely observed outside Iceland, its sole breeding area in Europe. Nests along streams and beside lakes in the lava regions, which afford suitable hollows. Male in breeding plumage distinguished from Goldeneye by *crescent-shaped white loral patch, which reaches above the eye*, and by fact that the *black on the back* is more extensive and *reaches far down onto the sides of the breast*. Also clearly *larger* than Goldeneye. More difficult to discern is the different head shape (more drawn-out in length, steeper forehead, peak less obvious and situated well forward on the crown), that the bill is slightly shorter and that the head has a violet, not green gloss. Note that young male Goldeneyes can have brownish-black heads with, depending on moult, quite noticeable crescent-shaped loral patches. When displaying, the male stretches neck forward with bill directed upwards and open (silent!). Wing noise deeper and harsher than that of Goldeneye, not so pure and musical. Females and immature males are difficult to distinguish from Goldeneye in the field. The female's bill is sometimes all-yellow (female Goldeneye: yellow band at bill tip in winter and spring). **V**

Barrow's Goldeneye

Harlequin Duck

Harlequin Duck *Histrionicus histrionicus* L 40, W 65. Like Barrow's Goldeneye an Icelandic (and American) species, rarely seen in the rest of Europe. Nests on islands in fast-flowing water-courses. In winter resorts to heavy surf off the rocky coast. Male easily identified by the light and dark pattern (at a distance the blue and red colours appear all-dark), the small size and the long tail, which is usually held slightly cocked. Female is smaller and darker than female Goldeneye, lacks white wing panels and has *three prominent white patches on the head*. Rarely mixes with other ducks. *Often swims with head nodding* in pace with leg strokes; sits high in water. Flight swift, actively pitches to and fro. Flies with very quick wingbeats (short-winged!) and head held high. Usually silent. The male sometimes gives a soft whistle, the female an agitated nasal jarring sound. **V**

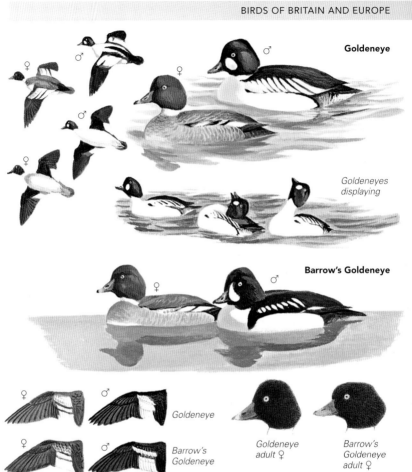

Goldeneye

♂

♀

♂

♀

♂

Goldeneyes displaying

Barrow's Goldeneye

♀

♂

♀ ♂ *Goldeneye*

♀ ♂ *Barrow's Goldeneye*

Goldeneye adult ♀

Barrow's Goldeneye adult ♀

Harlequin Duck

♂

♂

♀

♂

♀

♂

♂

41

Long-tailed Duck

Long-tailed Duck *Clangula hyemalis* L male 55, female 40, W 78. Abundant breeder on small lakes on the arctic tundra. Winters mainly in Baltic, also in the North Sea, mostly out of sight from land. Exposed to oil pollution. Active, apt to make pursuit flights in groups in wide sweeps over the sea. Flight swift and smooth, swinging from side to side. Wings (all-dark) somewhat swept back. Flying male from behind resembles Razorbill (white sides, dark centre). Alights with a big splash. In spring gathers closer to land in large, dense flocks which seethe with activity. The males stretch their necks up, raise their long tails (at other times often held trailing in the water) and sing 'ow-**ow**-owde**lee**' in chorus; nasal, wailing, far-carrying (reminiscent of distant bagpipes). The males moult in April from white-necked winter dress to predominantly blackish-brown summer plumage. Spectacular mass departure from Gulf of Finland towards tundra on some evenings around 20 May: the sky is full of large Vs which travel at high altitude across land for night stage of journey, singing and calling 'gack, gack,...'. **WP**

Velvet Scoter

Velvet Scoter *Melanitta fusca* L 55, W 92. Breeds mostly in taiga zone but also well above tree line; also in Baltic archipelagos. Return very late to breeding grounds. In winter marine like Common Scoter but keeps nearer to land and in smaller flocks. Males migrate to Danish waters as early as July to moult. Clearly larger than Common Scoter (half-way between it and Eider), tail shorter, forehead more sloping. The orange on male's bill is striking even at a distance, but white spot at eye is not and white secondaries often hidden. Female's whitish face patches very variable; the front one especially can be absent in old birds. Younger females often have whitish belly. The large white wing patches conspicuous in flight, make wings look narrow when above horizon, at distance. Looks heavier in flight than Common Scoter and holds head lower. Usually migrates in well-ordered bands of moderate size. When the pair circles over the breeding grounds the female calls a coarse, jolting 'pa-a-ah', otherwise silent. **WP**

Surf Scoter juv

Surf Scoter *Melanitta perspicillata* L 50, W 85. Irregular visitor from arctic America. Male unmistakable: has *large, brightly coloured bill* and *white patches on nape* (largest) and forehead. Female usually has two whitish spots on cheek as female Velvet (one or both may be missing) but sometimes also a smaller one on nape. Moreover, head profile is different, *bill being deep and wedge-shaped*, feathering reaching down onto bill, forehead practically missing. Equivalent of male's *black patch on base of bill* (sides) *is discernible.* Juvenile similar to adult female but belly usually much paler, and bill lacks black patch; also never has pale spot on nape. In flight Common Scoter is main confusion risk owing to size and *all-dark wings* (though male does not have brown primaries as male Common Scoter). Male has sonorous wing noise. **V**

Common Scoter ♀

Common Scoter *Melanitta nigra* L 50, W 85. Rare breeder in Scotland and Ireland, abundant breeder on lakes of N taiga zone, also on tundra. Large-scale migration by males in July to moulting waters west of Denmark. Females and juveniles follow in late autumn. Winters in large dense flocks (rafts) along N Atlantic coasts, well out to sea but in shallow waters. Exposed to oil pollution. Male looks all-black when swimming but shows *medium-brown primaries* in flight (striking in sunlight). Female dark brown with *dark cap* and *pale cheeks*. Juvenile even paler on cheeks, also pale on belly. Rounder head and longer tail (often raised) than Velvet Scoter and characteristically *keep very close together* also when in quite small parties. Migratory flocks often huge, fly low over the sea in scythe-shaped formation, side projection of which looks like a toy kite: crowded 'head', long snaking 'string'. Spring migration overland by night at great height, announced by call of males: a mellow, piping 'pew, pew, pew, . . .', slurred but amazingly far-carrying. Male has whistling wing noise, but only at take-off. **RWP**

Common Scoter

summer

♀

♂

summer

♂

♀

Long-tailed Duck

♀

♂

winter

♂

winter

♀

♀

♂

Velvet Scoter

♀

♂

♀

♂

♀

♂

Surf Scoter

♀

♂

♀

♂

♀

♂

Common Scoter

♀

♂

43

Eider

Common Eider ♀

Eider *Somateria mollissima* L 60, W 100. Breeds abundantly along coasts of N Europe, mostly resident in areas where ice allows, migrant (huge migratory movements) in Arctic Ocean and Baltic Sea. A big and heavy diving duck. *Male for the most part white* (in eclipse in late summer dark brown with white forewing), *female uniformly brown-mottled. Bill wedge-shaped and pointed,* pale greyish. Juveniles darker than adult female, young male darkest, has pale supercilium. *Heavier wingbeats* than other ducks, holds the *head low* in flight. Courtship in flocks, when the males prance and utter far-carrying, deep 'a-**ooh**-e' calls, while the females give a clucking 'kok-ok-ok-…', like a throbbing motor engine, which often forms dense sound effects during calm spring nights. Sitting female very confident. Clutch insulated by wreath of famous Eider down. The young broods hatch early, join up together and are attended collectively. The Baltic males gather in large flocks at sea from the end of May, after which they move along with younger birds to Danish waters and North Sea coast, to moult in July and Aug. First-summer males are brown-black with white breast. **RW**

King Eider

King Eider ♀

King Eider *Somateria spectabilis* L 55, W 92. Common in arctic regions, where like Long-tailed Duck nests mainly beside tundra meres. Otherwise maritime. Rare visitor south of Arctic Circle. Behaviour more or less as Eider, with which stragglers seek company. When occurring in numbers always form clean, dense flocks, usually further out at sea than Eider. Adult male unmistakable, has a *deep sideways-flattened, orange-red bill shield* (though this shrinks back in July, when also a rather drab dark brown eclipse plumage is assumed). In flight *black back* and *large white wing panels* are visible at long range. When swimming, shows two black fins on the back. Female resembles female Eider, but is clearly *smaller* and has a more 'conventional' shape of the head (quite different from the wedge-shaped head of Eider) due to considerably *smaller and thinner bill.* Moreover, *bill is dark,* and its *dark gape-corners stand out well against pale bill-base surround,* giving a strikingly 'happy' look. Pale upper eye-lid and *U-shape of flank markings* can be seen at close range. Immatures during their first winter are treacherously similar to Eider in plumage, greyish-brown and without U-shaped markings; identified only by *small bill* and by the *feather abutment.* On male King Eiders in obscure plumage, the *rounded nape* should be noted and also that bill is pink, not dirty yellow or grey. **V**

Steller's Eider *Polysticta stelleri* L 45, W 75. Arctic species, in W Europe rare winter vagrant. Seems not to be at home on open sea, often stays close in beneath cliffs, enters shallow rocky shore waters and feeds by up-ending, but certainly dives too in deeper water. Size as Goldeneye. Male unmistakable (in eclipse plumage, however, drab dark brown apart from white wing panels). Female difficult to identify; has much of a small Eider about her in plumage pattern and behaviour, but the head (rather large) has a totally characteristic profile with *rather long bill,* low but convex forehead, *flat crown* and *sharply angular crown/nape.* A diffuse pale ring around the eye visible at least at certain angles. *The speculum is framed by conspicuous white edges* (otherwise no features in common with Mallard). Tips of the downcurved elongated tertials in the adult female are usually pale, forming then a *pale transverse pattern* at stern. Immature *lacks* this character, is *further usually slightly paler brown,* especially on belly but the difference is not always easy to see. Furthermore, immatures *lack blue speculum* (have dull brownish-grey one) and have *distinctly narrower white bands framing the speculum* than in adult female. Young male has dark-shaded chin, forehead and nape. Flight like Common Scoter's. Pronounced wing noise, like something between Goldeneye and Mallard. **V**

Steller's Eider

imm. ♂

Eider

♀

♂

♂

♀

♂

King Eider

imm. ♂

♂

♀

♂

♂

Steller's Eider

♂

♀

♂

♀

♀

♂

45

Sawbills (subfamily Merginae)
Fish-eating, diving waterfowl with saw-toothed edges to the bill. All have white wing panels.

Red-breasted Merganser

Red-breasted Merganser ♀

Red-breasted Merganser *Mergus serrator* L 55, W 85. Breeds fairly commonly in N Europe along coasts but also beside larger clear inland waters and rivers, particularly in upland regions. Nest is on the ground, under bushes. Late breeder. In winter almost exclusively coastal. The male is characteristic (but moulting and immature males with brown head like female are a common sight). Female resembles female Goosander but distinguished by *darker*, more brown-toned grey *back, paler brown head* (cinnamon-brown rather than chestnut-brown), *thinner and more pointed crest* (not thick and drooping), less sharply contrasted pale chin, more grey-spotted breast and above all *much less sharp division between brown neck and grey body. The white speculum patch is divided by a narrow dark stripe.* In flight appears to have more slender head. The male displays with strange curtseying body movements, squeezing out nasal 'eh'. The female calls 'prrak' in circle flight over land. **RW**

Goosander

Goosander ♀

Goosander *Mergus merganser* L 64, W 95. More widespread than Red-breasted Merganser, fairly common on clear waters, also on coast in far north. Nests in tree holes and nestboxes. In winter on open lakes and reservoirs. Male in winter has splendid *salmon-pink underparts*, fading to white in Apr (the same applies to female's belly). Female's features compared with rather similar female Red-breasted Merganser are: greyer back, *darker, reddish-brown on head*, thick drooping nuchal crest, *sharply contrasted pale chin* and *sharp division between brown neck and grey body. The white speculum is not divided* (fig. left). Displaying males in winter utter a murmuring frog-like 'oorrp, oorrp…'. In spring a related more penetrating, metallic ringing 'drruu-drro' is heard. The female utters a 'skrrak, skrrak…' in flight. Fenno-Scandian males fly to N Norway for wing moult. In autumn gathers in N Europe in thousands at favoured lakes, fishes collectively in driving cordons, eagerly attended by gulls. **RW**

Smew

Smew *Mergellus albellus* L 40, W 65. Breeds in the northern taiga in tree holes and nestboxes beside small lakes. Male characteristic (but does not lose female-like eclipse plumage until Nov). Female and juvenile distinguished by *white cheeks*. Rests on lowland lakes and on coasts, often with Goldeneye and Tufted Ducks, and fishes in shallow water. In winter on reservoirs, lakes, occasionally sheltered bays. Poor flock unity, inclined to make aerial excursions. Flight swift and agile. The male displays by raising the crest on his forehead and drawing back his head on to the back, though with bill still pointing forward. **W**

Stifftails (subfamily Oxyurinae)
Small with short, thick necks and large bills. Tails markedly long and stiff, often kept cocked.

White-headed Duck

Ruddy Duck

♂

White-headed Duck *Oxyura leucocephala* L 46. Rare breeder in S Europe on freshwater swamps and brackish lagoons. Male has strikingly large and *heavy bill*, a beautiful *pale blue* in colour. The head is largely white, with black markings only on crown and hind nape, occasionally also chin. The neck is black, and *the body* if anything *chestnut-brown*. Recognised in flight by *short, rounded uniform-coloured wings* and *long tail*. Female told from female Ruddy Duck by a *shade deeper bill* and *more distinctly marked dark band on cheek*. Rarely takes wing, dives all the more frequently.

Ruddy Duck *Oxyura jamaicensis* L 41. This American species breeds locally in England, originally escaped from captivity. Male reddish-brown with white *undertail-coverts*. Note head markings and bill shape. Female is distinguished from female White-headed Duck by *weaker bill* and *only* faint *dark cheek bar*. **R**

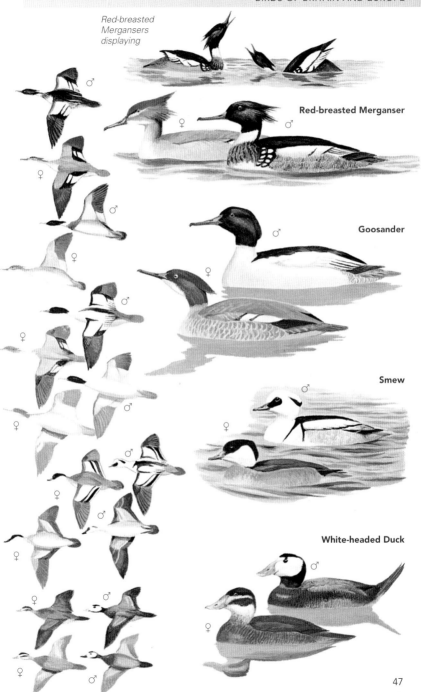

Red-breasted Mergansers displaying

♂

♀

Red-breasted Merganser

♀ ♂

♂

♀

Goosander

♂

♀

♂

♀

♂

Smew

♂

♀

♂

♀

♂

White-headed Duck

♀ ♂

♀

♂

♀ ♂

♀ ♂

47

GALLINACEOUS BIRDS (order Galliformes)

Togerther with waterfowl the oldest, most basal group of birds in Europe. Terrestrial. Have plump bodies and short bills. Short, broad, stiff, bowed wings. Have strong feet and can run quickly. Fly with series of rapid wingbeats alternating with long glides.

Grouse (family Tetraonidae)

Medium-sized or large. Nostrils and feet feathered. Several species have elaborate courtship displays. 5–12 eggs.

Willow Grouse

Willow Grouse *Lagopus lagopus* L 40. Numbers vary periodically, but quite common in taiga zone near boggy terrain and, most abundantly, in mountains, mainly in birch forest near overgrown brooks but also in damp willow sections on otherwise barren hillsides. In winter in small flocks in birch and coniferous woods of valleys. In the *white winter plumage* both sexes resemble Ptarmigan (though *never has black loral streak*), but altitude found usually distinguishes the two. During summer, male especially, recognised by *chestnut-brown elements* in plumage. In late spring, male is white with head, neck and upper breast deep reddish-brown; in summer and autumn plumage is mottled reddish-brown with belly and wings white, resembling the hen, which is quicker to moult the winter plumage. Grouse fly in characteristic manner of game birds: rapid wing-beats alternating with long glides on rigid, bowed, slightly depressed wings. Cock regularly bursts into barking laugh (far-carrying, nasal, choking) when flushed: '**keh**-uk, **keh**-hehehehehe-e**heh**-**eheh**, **eheh**'. In spring cocks, often many gathered loosely together, display at night, uttering laugh described above as well as ventriloquial 'go-**back**, go-**back**' and evenly accelerating 'ka, ke-ke-ke-ke-kekekeke**kerrr**'. Short aerial excursions (flutters up, glides away and down) are an integral part of the display. The cock will approach if the hen's 'nyow' call is imitated. Silent in winter. Cock takes part in care of chicks. Chicks are able to fly when still small (as all chicks of grouse family).

Red Grouse

Red Grouse *Lagopus l. scotica* and *L. l. hibernica* L 38. The British and Irish Red Grouse are two races of the Willow Grouse very closely related to each other and not separable in the field. Previously considered together as a separate species. Inhabit upland heather moors. Throughout the year chestnut-brown (hen less reddish) with *dark wings* (darker than hen Black Grouse). Calls same as Willow Grouse. **R**

Ptarmigan *Lagopus muta* L 35. Breeds in higher mountains in numbers that vary periodically. Less abundant than Willow/Red Grouse. In summer lives high up in lichen region, often near boulder ridges, i.e. as neighbour of the Snow Buntings, above the Dotterels. Winters, sometimes in large flocks, lower down on the bare mountain, often in upper birch forest, where Willow Grouse also found. Differentiated from latter in *white winter plumage* by *black loral streak in male*, sometimes an ill-defined one also in female. Call, when uttered, is species-specific. In late spring male becomes *blackish-grey on head/neck*, in early summer gets blackish and brown-grey over whole back, then in late summer acquires autumn plumage, which is more *grey-blue vermiculated*. Therefore very different from reddish-brown male Willow/Red Grouse. To separate the drab grey-brown (wings of course white) hen from hen Willow Grouse is much more difficult, but ground colour is more buffish-yellow, not so rusty-red; altitude where found a guide. Cock's call *very hard crackling*, belching 'arr, arr-**arr**', very different from laughter of Willow/Red Grouse. Usually silent when flushed, though. Flight very swift. Display of cock includes flights ending in steep ascent and gliding descent (reminiscent of a clay pigeon). **R**

Ptarmigan

Willow Grouse

early summer ♂

autumn ♂

winter ♂

autumn

♀ *summer*

winter

Red Grouse

summer ♀

summer ♂

♂ *summer*

autumn ♂

Ptarmigan

summer ♀

winter ♀

early summer ♂

autumn ♂

♂ *winter*

winter ♂

49

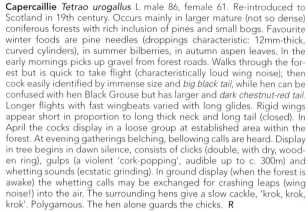

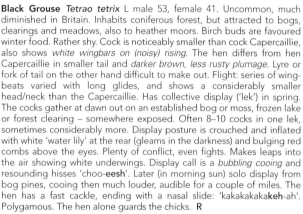

Capercaillie *Tetrao urogallus* L male 86, female 61. Re-introduced to Scotland in 19th century. Occurs mainly in larger mature (not so dense) coniferous forests with rich inclusion of pines and small bogs. Favourite winter foods are pine needles (droppings characteristic: 12mm-thick, curved cylinders), in summer bilberries, in autumn aspen leaves. In the early mornings picks up gravel from forest roads. Walks through the forest but is quick to take flight (characteristically loud wing noise); then cock easily identified by immense size and *big black tail*, while hen can be confused with hen Black Grouse but has larger and *dark chestnut-red tail*. Longer flights with fast wingbeats varied with long glides. Rigid wings appear short in proportion to long thick neck and long tail (closed). In April the cocks display in a loose group at established area within the forest. At evening gatherings belching, bellowing calls are heard. Display in tree begins in dawn silence, consists of clicks (double, with dry, wooden ring), gulps (a violent 'cork-popping', audible up to c. 300m) and whetting sounds (ecstatic grinding). In ground display (when the forest is awake) the whetting calls may be exchanged for crashing leaps (wing noise!) into the air. The surrounding hens give a slow cackle, 'krok, krok, krok'. Polygamous. The hen alone guards the chicks. **R**

Capercaillie

Black Grouse *Tetrao tetrix* L male 53, female 41. Uncommon, much diminished in Britain. Inhabits coniferous forest, but attracted to bogs, clearings and meadows, also to heather moors. Birch buds are favoured winter food. Rather shy. Cock is noticeably smaller than cock Capercaillie, also shows *white wingbars on (noisy) rising*. The hen differs from hen Capercaillie in smaller tail and *darker brown, less rusty plumage*. Lyre or fork of tail on the other hand difficult to make out. Flight: series of wingbeats varied with long glides, and shows a considerably smaller head/neck than the Capercaillie. Has collective display ('lek') in spring. The cocks gather at dawn out on an established bog or moss, frozen lake or forest clearing – somewhere exposed. Often 8–10 cocks in one lek, sometimes considerably more. Display posture is crouched and inflated with white 'water lily' at the rear (gleams in the darkness) and bulging red combs above the eyes. Plenty of conflict, even fights. Makes leaps into the air showing white underwings. Display call is a *bubbling cooing* and resounding hisses 'choo-**eesh**'. Later (in morning sun) solo display from bog pines, cooing then much louder, audible for a couple of miles. The hen has a fast cackle, ending with a nasal slide: 'kakakakaka**keh**-ah'. Polygamous. The hen alone guards the chicks. **R**

Black Grouse

'Rackelhahn' Hybrid between hen Capercaillie and cock Black Grouse. Well known in Fenno-Scandia from time when gamebird populations were strong but cock Capercaillies were severely reduced through shooting displaying birds. The males emerge on Black Grouse leks, are aggresive and disturb. Has 'shrunken' cock Capercaillie tail and head like cock Black Grouse. Overall size intermediate.

'Rackelhahn'

Hazel Grouse *Bonasa bonasia* L 35. Sedentary and local in central and eastern half of Europe in coniferous forests, preferably damp, dense and tangled spruce with birch and alder beside the streams. Occur in pairs. Difficult to see but not shy, can be called up by whistling its call. Rather *greyish-brown*, the sexes relatively similar, though hen's throat brown, smudgily bordered white, cock's brownish-black with more distinctly defined white border. *Crown tuft*, twitches when nervous. On rising, lower back and tail appear uniform ash-grey. Characteristic noise from series of wingbeats: 'boorr, boorr'. Usually lands in trees. Advertising call thin like Goldcrest's but drawn-out, sucking: 'tseeuu-**eee** titititi'. Alarm a very rapid twitter, 'pyittittittittitt-ett-ett'. Chicks only a few days old can fly up into the trees, distinguished by dark line through eye; tended by the hen alone.

Hazel Grouse

Capercaillie

♀

♂

♂

♀

chick

♂

courtship
display

Black Grouse

chick

♂

♀

♂

♀

males displaying

giving
advertising
call

♂

♀

♂

Hazel Grouse

chick

51

Partridges and pheasants (family Phasianidae)

Partridges and pheasants live in open country such as arable land, heaths or sunny mountain slopes. Often run, rather reluctant to take flight. Large clutches.

Chukar *Alectoris chukar* L 33. Breeds in open barren mountainous country, in Europe in Thrace (Greece) and bordering parts of Bulgaria. Very like Rock Partridge, but *bib is creamy-white* (not snow-white) and its black upper border touches only uppermost part of bill base. Call: nervous series of nasal, cracked clucks in falsetto, e.g. 'kakakakaka-chuck**ar**-chuck**ar**-chuck**ar**…'.

Chukar

Rock Partridge *Alectoris graeca* L 35. Scarce on rocky mountainsides in SE Europe, from W Alps to W Bulgaria, mostly at altitude 1200–1500 m. Population has declined markedly in many areas. Prefers sunny, southward-facing slopes with plenty of rocks and boulders and a mosaic of meadows and bush vegetation. Forest is often also accepted (in Balkan peninsula: deciduous forest) if conditions otherwise right. In winter often goes up to exposed windblown areas. Runs very ably (especially uphill), reluctant to fly. Very like Chukar but has *pure white bib*, and its black upper border often follows upper mandible down to gape-side. Moreover, chest is on average purer grey (more grey tinged buff in Chukar), and *whitish stripe above eye is thin*. Differs from Red-legged Partridge mainly in that *black lower border of bib is distinct*, not broken up in streaks on breast. Call: repeated series at galloping pace of quite deep and hollow clucks, e.g. '**che**kore-**che**kore…', accelerating and increasing in intensity towards end. When flushed whistles a shrill 'pyeet-pyeet-pyeet'.

Rock Partridge

Barbary Partridge *Alectoris barbara* L 33. Breeds on dry bushy mountainsides, in Europe on Sardinia (probably introduced) and Gibraltar (definitely so). *Blue-grey bib framed with chestnut-brown.* Brown crown/central nape shows up well on rising, resembles mohican hair-cut. Call: series of shrill, broken monosyllabic clucks with interposed double notes (trotting rhythm with 'stumbling steps'), e.g. 'krett krett krett kret**err** krett krett…'.

Barbary Partridge

Red-legged Partridge *Alectoris rufa* L 35. Common in agricultural country, on dry bushy heaths and also in rocky mountains. Introduced to Great Britain in 18th century. Like Grey Partridge, but small black-framed bib is striking. At closer range note that *lower area of the black frame is broken up into small black patches* (species-specific) and that bill and legs are gaudy red. On rising none of this is visible, Red-legged is then very like Grey Partridge with its rust-red tail. But even directly from behind one can see some *orange*, not present on Grey Partridge. Immature very like Grey Partridge, has not red legs (but greyish-pink, immature Grey brownish-yellow), but has diffuse dark vertical bars on flanks instead of yellowish-white lengthways streaks. More inclined to run away than the Grey; differs also in habit of perching on fence posts and even up trees. Call: rhythmically repeated notes in hoarse broken voice (rather like Grey Partridge), 'kuchek-**cher**-kuchek-**cher**…'. R

Red-legged Partridge

Grey Partridge *Perdix perdix* L 30. Commonest and most widespread of Europe's partridges. Found in open country, especially farmland with hedgerows. Numbers fluctuate. Completely terrestrial. Cock and hen together tend the chicks. The coveys keep very close together, squat firmly. All rise at same time with wing noise and loud 'grrree-grrree…'. Clearly smaller than Pheasant (but see p.54 about pitfall with half-grown Pheasant chick). Rather grey-brown, but rusty-red tail conspicuous on rising. *Head orange* (hen has pale eyebrow). Cock has large *dark brown patch below*, hen usually a less obvious patch. On spring evenings cock's creaky 'kierr-ik, kierr-ik' (= Perdix!) is heard. R

Grey Partridge

Chukar

Rock Partridge

Red-legged in flight

juv. Red-legged

Barbary Partridge

Red-legged Partridge

juv. ♀ ♂

Grey Partridge

♂

53

Pheasant

Pheasant *Phasianus colchicus* L male 85, female 60. Originally introduced from SW Asia. Common in open wooded terrain and agricultural land with copses, hedges, reeds etc. Often seen on open fields. Both sexes have *long, pointed tails* and short, rounded wings. The cock is very colour-ful, with bright red cheek patches contrasting with greenish-black head and neck. Plumage varies depending on the origin of the introduced stock. Usually distinct *white ring around the neck*. The hen is more nonde-script pale brown with dark markings but the long tail is characteristic. Note that half-grown but fully fledged Pheasant chick has red-tinted short tail, can therefore be confused with Grey Partridge. Takes flight in rapid noisy climb when put up. Flies rapidly, but only short distances. Spends the night in trees, often in small flocks. Feeds on spilled grain, seeds and berries. The cock's loud, explosive, two-note hacking call is followed by a series of noisy wingbeats. **R**

Quail

Quail *Coturnix coturnix* L 18. Formerly abundant, has lately recovered markedly after deep decline in 19th century. Still scarce in Britain, occur-rence fluctuating much. Long-distance migrant, arriving in May, returning in Oct. Inhabits large open fields of corn and grass, also clover pastures, keeps well concealed in the vegetation. Much *smaller* than other gallina-ceous birds: size of a barely half-grown Grey Partridge chick. Has *no rufous-red on the tail*. Is washed-out brown with paler streaking on back and sides. Adult male has variable amount of dark on throat (from whole throat dark to just some dark lines), unlike female. Very difficult to flush, runs cleverly away in cover. Flies low and markedly slowly, hunched up with retracted head, appears round-backed. Wings are surprisingly long and narrow (relatively speaking), held bowed with very fast, shallow beats. Glides between series of wingbeats almost negligible. In flight does not resemble a small Grey Partridge but rather a Snipe, albeit a straight-flying and low-flying one. Attracts attention mostly by its song, a far-carrying, tri-syllabic whistle like dripping water, '**kwic**, kwick-**ic**' (often rendered 'wet my lips'), which is repeated persistently. May be heard for a large part of the summer (June–Aug), both day and night but mostly at dusk. Also, the male has a peculiar grinding, subdued and slightly ventriloquial call used in anxiety or mild alarm, 'grrev-ev', repeated in much the same machine-like way as the song. **S**

BUTTON-QUAILS (order Gruiformes, family Turnicidae)

These are small, quail-like birds related to cranes and rails (which are else treated on p.106–113). Males take care of the young, and female has the more colourful plumage. (The single European species is treated here for easier comparison with the rather similar Quail.)

Small Button-quail

Small Button-quail (Andalusian Hemipode) *Turnix sylvaticus* L 16. One of the rarest and most enigmatic birds in Europe – if it in fact still breeds in S Iberia. Has its main distribution in sub-Saharan Africa and in SE Asia, but at least used to have also a small population on the Iberian S coast, notably in Andalucía (hence alternative name), and in NW Africa, where it has been found on grassy heaths in dry tussocky fields of dwarf palms and asphodels. Very shy and retiring and difficult to flush. Is a brown, small, ground-living bird like Quail, but differs from that on *pale orange-brown breast-patch* and *prominent brown-black spotting on sides*. Female is slightly bigger and neater than male. When seen in flight, note *smallness* and *very short tail*. Also, *upperwing-coverts are pale in contrast to flight-feathers* (unlike in Quail). Wings produce whirring noise. The song by the female is a characteristic muffled and deep hooting (like distant fog horn or lowing cow) 'hooh, hooh, hooh, hooh', heard particularly at dusk and dawn on clear nights.

♂

♀

Pheasant

♂

♀

♂

♀

Quail

♀

Small Button-quail

55

DIVERS (order Gaviiformes, family Gaviidae)
Divers (also called loons) are completely adapted to a life on and in the water. Powerful legs, placed far back, with webbed feet. The wingbeats are relatively fast, and the birds never use gliding flight. Over longer distances they fly high (often at heights of 20–70m), in contrast to grebes, which almost touch the tops of the waves. When diving, the divers disappear with a smooth, neat dip. They nest at the water's edge. Usually 2 eggs.

Red-throated Diver

Red-throated Diver *Gavia stellata* L 57. Breeds in N Europe and Scotland, typically at tarns on taiga bogs and tundra. Often flies long distances to larger lakes or the sea to fish. Winters mainly along coasts. When swimming often holds its *head and bill* (slender and upturned) *pointing markedly upwards*. In summer easily recognised by *reddish-brown throat*. In flight very like Black-throated but can often be told by *feet projecting less*, greater tendency towards *hunched back* and *sagging neck*, quicker wingbeats, higher upstroke, more backwards-angled wings and habit of *lifting head repeatedly*. Winter plumage paler than in Black-throated and with *more restricted grey on hindneck*, and *eye usually clearly white-framed*; back is sprinkled with small white spots; *side of body entirely dark above water line*. Male has continuously repeated, loud display call, 'oo rroo-u, oo rroo-u, oo rroo-u,…', accompanied by female's louder and shriller 'aar-roo-aarroo-aarroo-…'. Also has drawn-out wailing 'eeaaooh'. Most often heard call is rapid goose-like cackle in flight, 'gak-gak-gak-gak-…'. **RWP**

Black-throated Diver

Black-throated Diver *Gavia arctica* L 65. Breeds in N Europe and Scotland on lakes and lochs with deep, clear water and fish; rarely at coast. Migratory, winters along sea coasts. Character in summer are *black throat and chin* and pale grey crown. In winter plumage, back is dark grey; *on swimming birds a white patch is often visible on rear of body at the water line*. Bill held almost *horizontal* when swimming, is dagger-shaped, medium-heavy and straight, proportionately slimmer than in Great Northern Diver. Call on the breeding grounds (most often at night) is a desolate, mournful, far-carrying 'kloowee-kow-kloowee-kow-kloowee-kow-klowi'. Other calls: resounding 'aah-aw' like calling gull, and hard 'knarr-knorr'. Silent in flight. **RWP**

Great Northern Diver

Great Northern Diver *Gavia immer* L 75. Mainly a New World species. In Europe breeds on inland lakes in Iceland. Winters mainly along coasts of N and W Europe. Size rather variable, usually considerably larger than Black-throated Diver, and has proportionately much more powerful neck and bill. Immatures and winter plumage adults as a rule have quite pale, greyish-white bill (but *culmen and tip* always dark; mostly pale in White-billed). Winter plumage like Black-throated's but has *white eye-ring*, has crown and hindneck darker than back (converse in Black-throated) and often *broad dark half-collar* on lower neck. Wingbeats fairly composed and elastic; feet protrude far behind. On breeding waters gives loud screams and yodels ('maniacal laughter'). **W**

White-billed Diver

White-billed Diver *Gavia adamsii* L 80. Breeds in N Russia and Alaska. Winters chiefly in Norwegian waters. The largest diver, near enough identical to Great Northern but with *bill greyish-yellow-white, slightly upturned* and on average a shade longer. The culmen is completely straight in adults, in immatures (and in Great Northern Diver at all ages) usually convex. When swimming, holds the *bill pointing upwards* like Red-throated Diver. In summer plumage the white spots on back and neck are a shade larger and fewer than in Great Northern. In winter and immature plumages *the neck and the side of the head are paler* than in Great Northern and the *culmen and tip of bill pale*, at least on the outer part (rarely only the outer third pale). Like Great Northern, has a *dark half-collar* on lower neck. Wingbeats as slow as to recall Cormorant. **V**

head lift

Red-throated Diver

winter

juv.

Black-throated Diver

winter

Great Northern Diver

winter

White-billed Diver

winter

TUBENOSES (order Procellariiformes)
Tubenoses have external, tube-shaped nostrils from which excess salt is secreted. They are birds of the oceans and come ashore on remote islands and shores only to nest. Colonial nesters. Live on fish, plankton etc. Sexes alike. Utter calls only on the breeding grounds.
The tubenoses that visit European waters and coasts belong to the following families:

ALBATROSSES (family Diomedeidae), very large birds with long narrow wings and very powerful bills. Supremely skilled, indefatigable flyers, covering very long distances practically without any wingbeats, close to sea surface.

FULMARS, SHEARWATERS AND PETRELS (family Procellariidae), nearest in size to gulls. Small versions of the albatrosses, practising similar flight mode. Wings are held stretched, stiff shallow wingbeats are followed by long glides.

STORM PETRELS (family Hydrobatidae), small birds, barely larger than swallows. Lead a pelagic life like their larger relatives, feeding on open sea far from land, visiting nest burrows only at night. Flight is more 'erratic' and fluttering than that of the larger relatives.

Albatrosses
Albatrosses belong mainly to the southern hemisphere and visit Europe only as accidentals. Wingspan strikingly large. Although they may beat their wings in flight, for the most part they are seen gliding along following the contours of the waves on rigidly extended wings. Lay only one egg and do not breed every year. The Black-browed Albatross is the species most frequently seen in Europe.

Black-browed Albatross *Diomedea melanophris* L 80–95. W 213–246. Very rare, but the albatross one can most expect to see in Europe. Single individuals have on several occasions overflown to Scotland and the Faroe Islands and lived there for many years on bird cliffs among Gannets ('Solan Goose' in folk dialect). Very large with *long narrow wings*; wingspan is c. 125% of Gannet, c. 150% of Great Black-backed Gull. Appearance similar to the even rarer Yellow-nosed and Grey-headed Albatrosses but *entirely yellow bill, dark eyebrow streak* and fairly broad dark borders on underside of wing, *widest in front*, are characteristic features of the Black-browed. The immature has a grey crown, grey neck and dark bill. Most observations in Europe have been made during the summer months. **V**

Fulmars
Fulmars resemble the larger gulls in appearance and feeding methods, but unlike gulls also capable of shallow dive. Often nest on coastal cliffs, lay one egg.

Fulmar

Fulmar *Fulmarus glacialis* L 45, W 105. Nests in colonies on N Atlantic bird cliffs, increased markedly during the 1900s due to trawl-fishing. Nesting-cliff ledges are surrounded by greenery. Vomits stinking secretion over intruders at the nest. Between large and small gulls in size, but behaviour and proportions immediately distinguish it from these: flies like a miniature albatross, sails along in *long glides on stiff, straight wings*, exploiting the air currents close above the wave crests or alongside the nesting cliff faces. Wingbeats stiff. When swimming, floats high on the water. Leaps into flight when it takes off from the water. Has characteristically *robust head and neck area*, short tail and *short, thick bill*. Upperparts grey with *pale patches on primary bases*. Birds of southern populations (Britain, Norway, Iceland) gleaming white on head, neck and underparts, those of Arctic populations have these portions grey (pale to medium grey). Lacks white trailing edge to wing shown by gulls. Cackling calls can be heard from breeding site cliffs, and also from gatherings around offal dumps at sea. Follows ships. **RS**

Black-browed Albatross

adult

adult

adult

imm.

2nd-summer Gannet

Fulmar

pale adult moulting

'blue' phase

pale adult

59

Shearwaters (family Procellariidae)

Shearwaters have long, narrow wings, narrow tails and long, thin bills. They fly with a series of rapid wingbeats and long glides, usually near the surface of the water. The wings, typically stiff and slightly bowed downwards, are held low in gliding flight. In stronger winds the wing-beats are dispensed with; they glide along and rear up over wave tops. They live on small fish and crustaceans. Active at night at the breeding sites. Lay only one egg.

Macaronesian Shearwater *Puffinus baroli* L 28, W 63. Very rare visitor to European coasts, north to Denmark, from its nearest breeding range on Madeira and Canary Islands. Formerly known as 'Little Shearwater' (*P. assimilis*), this has now been split from other closely related forms in more distant waters, and the form *baroli* has become the species of European interest. In spite of being c. 20% smaller than Manx Shearwater it is still quite similar to this, sharing the same general plumage features of dark upperparts and white underparts. Note *weaker bill* of Macaronesian, and proportionately *slightly shorter and blunter wings*. Flight is fast with often *longer series of fluttering wingbeats* between *shorter glides* than on average in Manx, but note that wind has great influence on flight mode adopted. Most birds have paler tips to all or some upper wing-coverts, often forming a *paler panel along rear part of 'arm'*, not seen in Manx. On closer range note *whiter sides of head* in Macaronesian, with *dark eye surrounded by white* so that it stands out. Can *raise head in flight* rather like Red-throated Diver. **SV**

Manx Shearwater

Manx Shearwater *Puffinus puffinus* L 34, W 80. Common breeder in burrows on rocky islands and coasts in NW Europe, mainly in W and N Britain and in Ireland, locally also in France and Iceland. Forms large colonies at suitable sites, and European population estimated at c. 400,000 pairs. *Uniformly dark upperparts* contrast against very pale underparts, *whitish underwing narrowly bordered dark*. On closer range *dark can be seen to reach below eye* to cheeks, and there is often a *hint of a pale semi-collar* or indentation behind the dark cheeks. Flight is very swift, short series of stiff wingbeats alternating with long glides. Often seen in big flocks at sea. Does not follow ships. Calls from nest burrows at night in jerking rhythm, sounding 'weird', '**chi-ki** gah-ach'. **S**

Yelkouan Shearwater

Yelkouan Shearwater *Puffinus yelkouan* L 33, W 78. Mediterranean species breeding on coastal cliffs in burrows from Menorca to Aegean Sea. Known population c. 50,000 pairs but probably underestimated since thousands can pass the Bosporus in just an hour! Known to reach N Spanish coast in autumn. Very similar to Manx Shearwater, and separation between these usually rests heavily on locality. When seen well, Yelkouan appears more brown-grey on *upperparts, a trifle paler* than Manx (though can look quite dark on a cloudy day!), *less contrasting underwing* with greyish flight-feathers rather than black. Further, *feet protrude* on average more *beyond tail* (only a little at most in Manx), and there is *no hint of a pale semi-collar behind dark cheeks*. Can have a hint of a paler eye-ring. Calls said to resemble those of Balearic Shearwater.

Balearic Shearwater

Balearic Shearwater *Puffinus mauretanicus* L 38, W 86. Restricted as a breeder to Balearics in W Mediterranean, but regularly seen in N Atlantic from Morocco to Britain, occasionally even further north. Has declined recently and is now threatened, less than 2000 pairs remaining. Closely related to Manx and Yelkouan Shearwaters but is *slightly larger*, and often appears '*pot-bellied*' or a bit plump in shape. Short tail make *feet protrude somewhat in flight*. Grey-brown above, slightly *paler than Manx*, but greatest difference from this is *darker underparts*, with breast, flanks and lower belly/undertail-coverts being variably sullied brown-grey (in extreme cases even recalling Sooty Shearwater at long range!). Underwing less clean whitish, and often has *dark pattern on axillaries* in Great Shearwater-fashion. Calls at night from nest caves drawling, repeated '**aiiah**-eeech', first note in falsetto. **P/V**

Macaronesian Shearwater

moulting Aug. (fresh coverts)

worn sping/summer

Manx Shearwater

in low sun

foraging

Yelkouan Shearwater

Yelkouan *Balearic* *Yelkouan*

Balearic Shearwater

dark *typical* *typical* *pale*

61

Great Shearwater

Great Shearwater *Puffinus gravis* L 48, W 115. A large Atlantic shearwater with breeding period Nov–Apr on the Tristan da Cunha group of islands in S Atlantic (about halfway between South Africa and Argentine). In May–Jun migrates north to W Atlantic, feeding on banks outside N America, and in Aug–Oct southwards through eastern N Atlantic, possible to see in thousands in e.g. Ireland and Madeira. Compared with Cory's Shearwater, the most common alternative, Great Shearwater flies most of the time *faster* and in *shallower arcs, wings being held more straight*, and wingbeats (in series) are quicker. Contrast between *brownish-black cap* and *white cheeks* visible at quite long range, and the cap is further enhanced by hint of a paler nape collar. A *whitish 'U' on uppertail-coverts* can also be seen far. Medium brown back and innerwing contrast against dark brown outerwing. *Underwing* not quite as clean white as in Cory's, axillaries and inner coverts often *showing a dark diagonal pattern*. The *dark belly patch* is diagnostic but surprisingly difficult to see. **S**

Cory's Shearwater

Cory's Shearwater *Calonectris diomedea* L 50, W 118. Breeds in caves on rocky islands in the Mediterranean (race *diomedea*; but also one colony in the Biscay) and outside N Africa (race *borealis*; but one colony known also on Spanish SE coast). Appears Aug–Nov in the N Atlantic. Largest of the Atlantic shearwaters. Usually appears *rather uniformly grey-brown above*, with *brown sides of chest*. Some birds have a small amount of white on the uppertail-coverts like Great Shearwater, but never have blackish cap or pale nape collar; also, *underwings are unmarked all white*, including armpit, apart from thin brown margins. When seen well, races (or possibly separate species?) can often be separated: *diomedea* has whiter underwings, with white reaching far out towards tips of primaries, whereas *borealis* has much darker tips to primaries below. Flight slower and more leisurely than in other comparable species (though can be fast enough in a strong tailwind), and wings are kept more arched and angled back. Can look quite gull-like when it practices active flight on a calm day. However, series of relaxed wingbeats are characteristically interspersed with *long, straight glides close over sea*. Again, can look gull-like at a distance when resting on sea, but forms much denser rafts than any gull (more like e.g. Common Scoters). The only Atlantic shearwater that can be seen flying high up and even practice soaring. Sometimes follows ships, and often dolphins and schools of other small whales. At night a cacophony of weird calls (rather Kittiwake-like) can be heard from colonies. **P**

Sooty Shearwater

Sooty Shearwater *Puffinus griseus* L 45, W 105. Breeds in the southern hemisphere, south of 30°S, in burrows on islands. Occurs on the Atlantic Jul–Feb, most abundantly in Aug–Nov, also annually in the North Sea. More tied to coastal waters than most shearwaters. Easily recognised by its *uniform dark greyish-brown plumage* with poorly defined *pale band under the wing* (at very long distance looks all-dark). Still, can be mixed up with darkest extreme of Balearic Shearwater, but differs from that on *longer and narrower wings* and *faster flight*. Wingbeats faster even than in Great Shearwater, with which it often occurs in W Europe. Wings longer and narrower than in Manx Shearwater. In flight, normally *holds wings more backswept* than other *Puffinus* species. Always keeps very high speed – is something of the grey-hound of the sea. And can cover huge distances: one bird with transmitter flew 900km in one day, and annual minimum movements must amount to more than 65,000km! Follows ships only exceptionally. Often dives for food. **P**

Sooty Shearwater

*Balearic
Shearwater for
comparison*

Great Shearwater

borealis

**Cory's
Shearwater**

borealis

diomedea

Storm-petrels (family Hydrobatidae)

Small pelagic birds with fluttering, bouncing flight over the waves. Strong, hooked bills and tube-shaped nasal openings. Lay one egg in cave or under boulders.

British Storm-petrel

British Storm-petrel *Hydrobates pelagicus* L 15, W 37. Most widespread storm-petrel in Europe, breeding both in Atlantic and Mediterranean. *Smallest* storm-petrel with *rather blunt-tipped wings* which are often *held moderately angled back. Darker* than Leach's, lacking pale upper wing-panel (only has narrow trace), but having *prominent white band on under-wing.* Rump pure white, *tail square-cut.* Flight clearly distinct from that of Leach's, has *quicker wingbeats* and *lacking obvious shearwater-like arched glides.* Flight when feeding is 'erratic' and roving about, recalling a House Martin; can even tap feet on surface. Purposeful flight on migration a little like a small wader (Dunlin, phalarope). Calls at night at nest a drawn-out purring with inserted hick-ups. **RSP**

Wilson's Storm-petrel

Wilson's Storm-petrel *Oceanites oceanicus* L 18, W 40. May be seen far out in the Atlantic, mainly Aug–Dec when returning to breeding sites in the S Atlantic. Dark brown plumage, with white rump, which extends down to the sides of the undertail-coverts. Rather indistinct pale wing patches above. *Long legs,* which project just beyond tail. When looking for food, *dancing flight over the water's surface with wings held high, pattering along the water. Remarkably long glides on outstretched straight wings,* close above water, is characteristic. Wingtips are comparatively rounded. Lacks British Storm-petrel's obvious white wing-bar below, has only diffuse paling effect. Often follows ships in loose flocks. **V**

Madeiran Storm-petrel *Oceanodroma castro* L 20, W 45. Very rare visitor from breeding sites on Atlantic islands (one breeding site also off coast of Portugal). Slightly larger and longer-winged than British Storm-petrel, has more shearwater-like flight, lacks white on underwing. Does not follow ships. **V**

Leach's Storm-petrel

Leach's Storm-petrel *Oceanodroma leucorhoa* L 22, W 48. Widely distributed in N Atlantic and N Pacific, in Europe with scattered breeding sites in Scotland, Ireland, Norway and Iceland (largest population). Dark with prominent *white rump* and *pale grey wing panels above.* Has *forked tail.* The white *rump* has a variably obvious *central divide* (a hint only, or dark and broad making whole rump look dark). *Wings rather long and pointed,* usually carried with marked *angling at carpal.* Flight clearly different from that of British Storm-petrel: the *longer wings* are beaten at a slower pace (can recall a small Black Tern!), and *careening flight in shearwater fashion* is practised a lot. At times halts to tap the water's surface with its feet. Does not follow ships. Calls at night at nest a fast, rattling cooing ending with rhythmic crescendo. **SP**

Petrels (family Procellariidae)

Like small, robust shearwaters, with shorter and heavier bills than these.

Bulwer's Petrel *Bulweria bulwerii* L 28, W 70. Very rare visitor (incl. in Mediterranean) from Madeira and Canary Islands. Feeds far out to sea. Roughly the size of a tern. Appears *all black* at long range, but brown tinge and slightly paler innerwing panels visible when close. *Slender with long and narrow wings,* and *extended tail.* Flight comparatively active and roving, wingbeats rather quick. Careening flight in short, shallow arcs. **V**

Fea's Petrel *Pterodroma feae* L 35, W 88. Very rare visitor, breeds in small numbers e.g. off Madeira. *Very long and pointed wings. Underwing almost all-dark,* contrasting against *white belly.* Upperparts often appear uniformly grey (blackish 'W' difficult to discern) apart from *light uppertail.* Fast careening flight includes extremely bold and high arcs. **V**

Wilson's Storm-petrel

feeding

feeding

feeding

British Storm-petrel

Madeiran Storm-petrel

feeding

feeding

Leach's Storm-petrel

Bulwer's Petrel

Fea's Petrel

65

PELICANS AND ALLIES

PELICANS AND ALLIES (order Pelecaniformes)

Large, aquatic and fish-eating birds with all four toes connected by webs (paddle-footed). Most species breed in large colonies and are silent outside the breeding season.

GANNETS (family Sulidae) are represented in Europe by the Northern Gannet, which dives vertically for fish like a giant tern.

PELICANS (family Pelecanidae) have enormous bills with which they rake in fish, driving them by swimming in cordons.

CORMORANTS (family Phalacrocoracidae) dive from the surface and swim under the water. They are often seen perched on posts and cliffs with their wings extended.

Gannet

Gannet *Morus bassanus* L 92, W 175. Breeds on almost inaccessible precipitous rocky islands on Atlantic coasts, in immense colonies. Isolated colony on coast of NE England. Britain harbours the greater part of the world population, which has steadily increased over recent decades. Winters at sea. Very occasionally blown inland after severe autumn/winter storms. *Long pointed wings,* long wedge-shaped tail. *White with black wingtips* and *yellowish-buff head.* Juvenile is brown finely speckled white, acquires adult plumage by stages over 4–5 years. In the second autumn the head, belly and leading edges of the wings are pale. In the third autumn the first white secondaries appear interspersed among the remaining dark ones, and the crown and nape acquire the adult's yellowish tone. In the fourth autumn, dark central tail feathers and dark secondaries scattered among the new white ones are usually visible. Catches fish by vertical dives of up to c. 40 m height like a gigantic tern, but usually completely disappears under the water (folds back wings on entry). In feeding flight, *wingbeats are much quicker* than gulls, and glides are few. In strong winds often glides down between waves and up again, and immatures can then be confused with larger shearwaters. At breeding site makes loud gurgling calls. **RS**

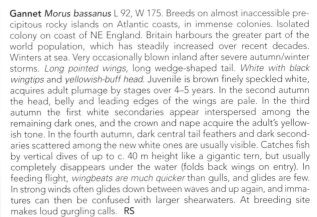

White Pelican

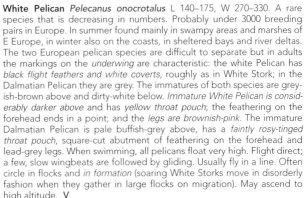

White

Dalmatian

White Pelican *Pelecanus onocrotalus* L 140–175, W 270–330. A rare species that is decreasing in numbers. Probably under 3000 breeding pairs in Europe. In summer found mainly in swampy areas and marshes of E Europe, in winter also on the coasts, in sheltered bays and river deltas. The two European pelican species are difficult to separate but in adults the markings on the *underwing* are characteristic: the white Pelican has *black flight feathers and white coverts,* roughly as in White Stork; in the Dalmatian Pelican they are grey. The immatures of both species are greyish-brown above and dirty-white below. *Immature White Pelican is considerably darker above* and has *yellow throat pouch;* the feathering on the forehead ends in a point; and the *legs are brownish-pink.* The immature Dalmatian Pelican is pale buffish-grey above, has a *faintly rosy-tinged throat pouch,* square-cut abutment of feathering on the forehead and lead-grey legs. When swimming, all pelicans float very high. Flight direct; a few, slow wingbeats are followed by gliding. Usually fly in a line. Often circle in flocks and *in formation* (soaring White Storks move in disorderly fashion when they gather in large flocks on migration). May ascend to high altitude. **V**

Dalmatian Pelican *Pelecanus crispus* L 160–180, W 310–345. Rare breeder on swampy ground and lakes in SE Europe. Now probably no more than 250 breeding pairs in Europe. In winter often moves to sheltered sea coasts. Distinguished from White Pelican by *wholly pale underwing.* Even at a distance the *body feathers* are seen to be *greyish-white* (not yellowish-rosy). The eyes are yellowish-white (not red) and the *nape feathers curly. The immature is considerably paler above* than immature White Pelican – see also under that species. Gregarious, as White Pelican.

Dalmatian Pelican

66

Gannet

adult

nesting colony

intemediate

juv.

adult White

White Pelican

Dalmatian Pelican

adult Dalmatian

juv. White

juv. Dalmatian

67

Cormorant

juv.

Cormorant *Phalacrocorax carbo* L 90, W 145. A widespread species, inhabits five continents. Nests colonially. Race *carbo* breeds along N Atlantic and Murman coasts, on rocky islands and cliff ledges (exceptionally in trees). Race *sinensis* breeds in central and S Europe (incl. Holland, Denmark, Sweden) in trees, often by fresh water, often in company of herons. The trees are killed by their droppings. In Britain mainly marine but avoids rough sea, prefers shallow coastal waters and estuaries, also visits reservoirs, rivers etc. *Large*, dark and reptile-like. In spring has white on chin and cheeks, a white patch on thigh and (for a short period) a varying amount of whitish hair-like plumes interspersed on hind parts of head. In *sinensis* these plumes are abundant and large, *much of head and upper neck* looking white in early spring. In autumn/winter almost all-dark; whitish on chin slight. Juveniles and immatures are brown-black, but *belly is whitish* with few exceptions (Atlantic juv. Shag: belly brown). All-dark birds can be difficult to tell from Shag, but *bill is heavier, head larger* and *more flat-crowned* and *angled at nape*. Swims low with neck erect, bill held up at an angle. Expert diver, dives with small jump or with a more graceful bow. Rests on rocky shores, on sandbanks (often many in a long line), piles, buoys etc, in upright posture, wings often typically held spread (to dry flight-feathers after dives). Flight with goose-like wingbeats, at times interrupted by short glides. Usually flies several metres above sea (cf. Shag). Large flocks fly in winding groups. Overflies land at great height, at which time also soars. Deep guttural noise at nest, otherwise silent. **RSW**

Shag

juv.

Shag *Phalacrocorax aristotelis* L 70, W 100. Breeds in colonies on rocky coasts. Habits much as Cormorant but at home also in rough sea and avoids fresh water, rests on cliffs, only rarely perches on piles, buoys etc. Adult all-black, glossed green, with *bright yellow gape*; in early spring also a recurved tuft on forecrown. Non-breeding Cormorant is similar (almost all-black too), but Shag has *slimmer neck, smaller and rounder head* with steeper forehead, *narrower bill* (note that occasional immature Cormorants have confusingly slender bill). Juveniles rather uniform brown below (juv. Cormorant: belly usually whitish) with well-marked whitish chin, but those of race *desmarestii* (Mediterranean and Black Sea) are extensively whitish below. Wing-coverts of juveniles and particularly of second-year birds are edged pale, giving *large pale wing panel in flight* (Cormorant: uniformly dark wings). Otherwise similar to Cormorant in flight; smaller size not obvious (and size of Cormorant varies a great deal), but *wingbeats noticeably faster*, slimmer *neck stretched* out (not slightly retracted and crooked), smaller head reaching upwards, *belly bulging*, all giving a somewhat tail-heavy look. Shag usually flies close to the water (Cormorant frequently higher up). **R**

Pygmy Cormorant *Phalacrocorax pygmeus* L 50, W 85. Breeds locally in SE Europe, in colonies in bushes beside lakes and rivers with large reedbeds, often together with egrets and herons. Often fishes in quite small rivers and pools out in the swamplands. Immediately distinguished from Cormorant and Shag by *small size* – is smaller than a Red-breasted Merganser – and in addition has different proportions: *smaller head* and in particular *shorter bill* (looks 'baby-faced'), *longer tail*. In breeding plumage head and neck are dark chestnut-brown, the body glossy greenish-black with small white feather tufts which stand out like white droplets (both sexes). The feather tufts are soon lost and the chin becomes whitish, the breast reddish-brown. Juvenile is dark brown with whitish chin and belly. Swims low in the water and perches to dry out like its larger relatives, but may also use reed stems and thin branches as perches. Flies with same wingbeat rate as Eider, with short glides interspersed; at long range and at poor angles, therefore, Glossy Ibis is a confusion risk.

Pygmy Cormorant

Cormorant

juv.

Atlantic race carbo *breeding*

Continental race sinensis *breeding*

flight formation

Atlantic adult non-breeding

Shag

juv.

non-breeding

breeding

juv.

non-breeding

juv.

Pygmy Cormorant

breeding

69

HERONS, STORKS AND IBISES (order Ciconiiformes)

Wading birds with long necks, legs and bills. Most live on smallish animals which they catch in shallow water. Some have long plumes (aigrettes) during the breeding season. Wings broad and rounded, tails short. Clutches 2–6 eggs.

Herons and their allies are divided into the following families:

HERONS AND BITTERNS (family Ardeidae), bills straight, flight slow with necks retracted. Most are colonial nesters. Partly nocturnal. Hoarse and muffled calls.

STORKS (family Ciconiidae), bills straight, flight with outstretched neck and slow powerful wingbeats. Soar readily. Plumages black and white.

IBISES AND SPOONBILLS (family Threskiornithidae), bills thin and curved or flat and spoon-shaped. Wingbeats fairly quick, necks outstretched.

Bittern

Bittern, camouflage posture

Bittern *Botaurus stellaris* L 75, W 130. Breeds sparingly in scattered pairs in large reedbeds. Polygamous: one male may have several females in the reeds. Partly diurnal but keeps well concealed. Clambers about, clutching bunches of reed stems. If alarmed, it stretches its bill straight up in the air (the 'bitterning' posture). Easiest to see on early mornings in summer, when it flies to and from fishing sites. In flight *retracted neck*, but the wingbeats are not sluggish and heavy like Grey Heron's but quick and even as in the smaller heron species. This, together with the *brownish-speckled appearance* and the ungainly shape, make it very owl-like in poor light. Immature Night Heron is a risk of confusion in southern Europe. The Bittern starts breeding activity early, in the north the first ones while the ice is still present, and the male's booming night-time call is uttered throughout spring and far into June; heard best at dusk and dawn. The powerful waves of sound, reminiscent of blowing into an empty bottle, audible over 5 km, are preceded by muffled intakes of breath; 'u u u u uh-**poh**, uh-**poombh**, uh-**poombh**, uh-**poombh**'. On dark autumn evenings far-carrying, hoarse 'kaau' calls are heard from flying Bitterns, at close range sounding like large gulls, at long range quite like the barking of a fox. **RW**

Little Bittern *Ixobrychus minutus* L 35, W 55. Shy and usually difficult to observe. Found in S and C Europe. Overshooting spring migrants regular further north (has bred in England). Inhabits dense vegetation in swampy areas, preferably large reedbeds, where it breeds in single pairs. Easy to identify by size and colour. In flight the contrast between the *pale wing panels* and the *dark wing and back* is clearly visible. Male shows more contrast than female, has *black back* and *brilliant buff-white wing panels*. Female is streaked brown on the back and has dirtier wing panels and *more streaked breast*. Immature is spotted brown like Bittern but also shows the adult's pale wing panel in flight. Sometimes conceals itself by standing motionless with bill pointing up as if paralysed ('freezing'). Flight characteristic with *Jay-like quick wingbeats and long glides*. Often flies very short distances, low over the reeds. Mating call is a quite muffled grunting 'grook' repeated rhythmically every two or three seconds and in very long series when the bird is in full song. Also has an excited, loud nasal 'kekekeke'. **SV**

Little Bittern

Bittern

*juv. Night Heron
for comparison*

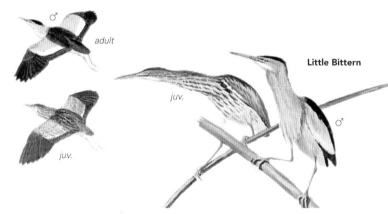

♂

adult

Little Bittern

juv.

♂

juv.

71

Little Egret

Little Egret *Egretta garzetta* L 60, W 92. Breeds in S Europe in colonies in marshes, river deltas and swamplands which have the necessary clumps of trees for nest-building. Regularly over-shoots in spring. During rest of year seen beside all kinds of shallow water, but especially in swamps/marshes with salt or brackish water. Very gregarious. Reliable character is the *yellow toes* contrasting with the all-black legs. *Bill entirely black.* (The Western Reef Heron *E. gularis*, a rare visitor from coasts of Africa and Red Sea, of which white form is extremely similar to Little Egret, is distinguished by shorter head plumes together with heavier bill, which turns brown or yellow outside breeding season.) Wingbeats in quick crow tempo. Shape and movements are more 'heron-like' than those of Cattle Egret and Squacco Heron. Captures its prey by standing in wait or by advancing slowly and stealthily. The elongated scapulars and nape feathers are worn only in summer. Great Egret is much larger, has differently coloured bill and legs, flies with more composed wingbeats and in flight legs project more than in Little Egret. Cattle Egret is more compact and has proportionately shorter wings. Call bubbling, frog-like 'gullagullagulla', also harsh raucous 'kark' calls. **WP**

Great Egret

Great Egret (Great White Egret) *Ardea alba* L 90, W 150. Scarce breeder in reeds in marshes, deltas and lagoons in SE Europe (has bred in Netherlands). During rest of year also found at other kinds of shallow water. Much larger than other white herons – almost the *size of Grey Heron*. Lores blue-green, gape extends far behind the eye. *Bill black with yellow base* (breeding period) or *all-yellow* (rest of year), dark toes and *reddish or yellowish-brown* (*breeding birds*) *tibia* separate it from Little Egret, which is also considerably smaller. Non-breeding birds have dark tibia, looking black-legged at distance. In flight the legs project farther behind tail than in Little Egret; wingbeats are also slower, as Grey Heron's. Like Little Egret, bears elongated scapulars, known as aigrettes, in summer plumage. The call is a harsh rolling 'krr-rr-rr-rra'. **V**

Squacco Heron

Squacco Heron *Ardeola ralloides* L 45, W 87. Breeds in S Europe in swamps and lagoons, where nesting takes place in reeds or trees, most often forming a minority in colonies of other small herons. *Pale ochre on body and neck* with contrasting *snow-white wings and tail* are characteristic. In the field the effect is that a standing Squacco looks mainly brownish but is transformed into an almost completely white bird when it flies off. Distinguished from Cattle Egret also by bill colour (largely blue with black tip when breeding, yellowish-green with black tip during rest of year). Often spends the day perched in trees or shrubbery and searches for food at dusk. Compared with Cattle Egret, is solitary, quiet and stealthy. Flight comparatively 'wobbly'. Call a harsh raucous 'krak', almost like Mallard. **V**

Cattle Egret

Cattle Egret *Bubulcus ibis* L 50, W 95. An expanding species, is seen occasionally in many places in Europe. Usually breeds in colonies in clumps of trees and bushes together with other small herons. Gregarious. Seeks food in fields and dryer marshland, usually in flocks, often alongside cattle, normally in considerably dryer terrain than other herons. At a distance the *plumage* appears *all-white*; in breeding plumage and at closer range it can be seen that the crown, breast and lower back have a yellowish-brown tone. At close range the leg and bill colour should be noted: legs yellow or reddish during breeding season, grey-brown or blackish at other times; bill yellowish (red-toned at pairing time). Has strongly 'undershot jaw', i.e. the *feathering on the lower mandible is conspicuous*. Migrates in long disorderly flocks at relatively low altitude. Flight silhouette in profile is markedly more distended, more *short-legged, short-billed and 'snub-nosed'* than Little Egret. Squacco Heron is smaller and flies more unsteadily, more like a Little Bittern. Calls quite subdued, slightly nasal, croaking, usually monosyllabic. **V**

Little Egret

nesting colony

non-breeding

breeding

Great Egret
summer

Squacco

Cattle

juv.

Squacco

Squacco Heron

adult
summer

juv.

Cattle Egret

adult
breeding

73

Grey Heron

juv.

Grey Heron *Ardea cinerea* L 95, W 185. The most abundant and most widespread of Europe's herons. Found on food-rich lakes, rivers and most other fresh waters, also on sea shores. Nests usually in large, noisy colonies in tall trees near water, but single nests not infrequent. Hardy, only retracts from snow and ice, needs open water for fishing. Stands motionless in shallow water in wait for fish, which it captures with a lightning-fast bill stab. The patient watching behaviour and the stiffly held and often retracted neck are very characteristic of herons in general. The Grey Heron is easily distinguished from other European herons by its size and the *grey, white and black plumage*. In flight the neck is always retracted; the heron then looks front-heavy. Over longer distances the Grey Heron flies at high altitudes, and can then be confused with large birds of prey because of its slow, heavy wing action. But even at long range the Grey Heron's particular characters in flight are obvious: *bowed wings* that *beat heavily and slowly*. Call a raucous 'kaark', often uttered by flying birds in the night. **RWP**

Purple Heron

Purple Heron *Ardea purpurea* L 85, W 135. Locally common on marshy land and swamps in S and central Europe. Spring migrants regularly overshoot. Nests in colonies, usually in reedbeds. Prefers denser vegetation than Grey Heron. The purplish-red elements in the plumage are difficult to see at a distance, when it looks generally *a little darker than the Grey Heron*. In flight the *forward-bulging crook of the retracted neck is less rounded than the Grey Heron*, forming a more pointed 'battering ram'. Head/neck more slender and snaky, bill more uniformly narrow, not dagger-shaped, and *toes are longer*, the hind toe sticking out more in flight. Adopts Bittern-like camouflage posture. **P**

Night Heron

Night Heron *Nycticorax nycticorax* L 60, W 112. Common in S and central Europe in swamps and marshes with fresh or salt water. Nests in clumps of trees in colonies with other small herons. Sturdy body and black, *grey and white pattern* distinguish adult. Gives rather pale general impression in flight. Juvenile is brown and can be confused with Bittern, but is *smaller* and has *prominent pale spots* on wing-coverts. In flight told on quicker wingbeats and darker plumage. Often spends the day in trees or bushes. At times seen searching for food in daytime but mostly at dawn and dusk. In flight the body is held slightly raised, the bill pointing slightly downwards; moreover the feet *do not form a rectangular blob* as in other small herons, *but a slender point*. Call a soft frog-like 'kooark'. **V**

Glossy Ibis

Glossy Ibis

Glossy Ibis *Plegadis falcinellus* L 60, W 90. Breeds locally in S Europe, colonially in marshes or trees by water. Has declined in 20th century, but recently a slight improvement. Feeds in marshland or on mudflats. Immatures may straggle far outside breeding range in Sep–Oct. The *curved bill* and at a distance the *all-dark plumage* are good field characteristics. At closer range reddish-brown on head and body, a shimmering green wing patch and narrow white bill base can be seen. In winter plumage, head and neck brownish-black, spangled with small white spots. Immature resembles winter adult, but whitish spots less well marked and back and upperwing duller and browner. Flies in flocks in long lines, wingbeats quick like a curlew's, interspersed with short stages of gliding, roughly as Pygmy Cormorant (possible confusion at long range). *The neck is extended in flight, legs protrude beyond tail-tip*. Calls are loud rumbling, belching and croaking. **V**

Sacred Ibis *Threskiornis aethiopicus* L 66. Widely distributed in sub-Saharan Africa. Has been introduced in SW France, and self-sustaining population now established there. *White plumage with black 'hind-bush'* (bushy tertials) together with dark neck and *dark head* make this species easy to recognise.

Grey Heron

adult

juv.

Purple Heron

adult

juv.

adult

roosting

*juv.
Night Heron*

Night Heron

adult

adult

Glossy Ibis

adult

*juv.
Glossy
Ibis*

adult

Sacred Ibis

juv.

75

White Stork

White Stork *Ciconia ciconia* L 110, W 180–220. Common except in northern part of breeding range, where it is decreasing in numbers. A bird of open country, preferring wet meadows and grasslands where it feeds on frogs, snakes, grasshoppers, fish, etc. Nests on roofs of houses, often on carriage wheels put up for the purpose, and in big solitary trees, sometimes in small colonies. Easy to approach, seldom shy. Walks slowly and in dignified manner. Easily distinguished from Black Stork by the *white upperparts*. Flies with straight neck, is often seen soaring high up in good thermals. Winters in Africa. The western population leaves Europe over Gibraltar, the eastern one (far and away the largest) by the Bosporus. It can be seen in immense soaring flocks over Istanbul at the end of Aug. These soaring flocks are characterised by their *teeming disorder*. Pelicans, which may also appear in huge soaring glistening white flocks on migration, may be confused but they always maintain a certain order, individual groups moving in synchrony and formation. The White Stork communicates with characteristic bill-clapping. **P**

Black Stork

Black Stork *Ciconia nigra* L 105, W 175–205. Rare. Inhabits wooded regions, usually by lakes, rivers and swampland surrounded by trees (nests in trees). Easily distinguished from White Stork by *black upperparts with metallic sheen*. At long range and against the light, the colour of the upperparts and the neck can be surprisingly difficult to judge owing to the effect of the glossiness. The immature is identified by greenish colour on legs and bill, not red. Usually, but not always, shyer than White Stork and because of its small numbers, rarely seen in company (except at the Bosporus in migrating flocks, at the end of Sept). In contrast to White Stork, has a strong voice, utters loud, shrill, raptor-like 'p(ee)luv, p(ee)luv'. Rarely claps its bill, however. **V**

Spoonbill

Spoonbill *Platalea leucorodia* L 88, W 130. Uncommon and with fragmented distribution. Found at shallow, open waters, reedy marshes and lagoons. Nests colonially in larger reedbeds, sometimes builds in trees and bushes. Distinguished on the ground from the white herons by the *broad and very long bill. Holds neck straight out in flight.* The long crest is worn only by adults in summer. Immature is black on tips of primaries. Flies in flocks, usually in a line. Flies with very *much faster wingbeats than the storks*, if anything more like the rhythm of the Cormorant. Also glides and soars. *Sweeps head/neck from side to side when seeking food* in shallow water. Usually silent, but bill-clapping may be heard from birds when excited. Occasionally utters sound resembling clearing of one's throat. **SP**

FLAMINGOES (order Phoenicopteriformes)

Very long legs and necks as well as heavy downward-bent bills. Nest in large colonies.

Flamingo

roseus

chilensis

Flamingo *Phoenicopterus roseus* L 135, W 155. In Europe breeds in a few colonies but each containing many individuals. Occasional individuals may be seen anywhere in Europe, but are in most cases escapes from zoos. Lives and breeds in colonies on mud pans and banks with shallow, salt water. At long range flocks on the ground look like white stripes, flocks in flight like rosy-coloured clouds. Flies in a line over longer distances. *Neck and legs extremely long, held slightly drooped in flight. Bill short, thick and bent. Immature is brownish-grey-white without any pink, has dark legs and dark bill.* Seeks food by skimming in mud in shallow water with bill upside-down. Has various goose-like trumpeting and cackling calls, often given in flight. (Zoo escapes are often shown to be so by belonging to one of the two American species, American Flamingo *Ph. ruber* or closely related *Ph. chilensis: ruber* is strongly rosy-pink over the whole of its plumage; *chilensis* is appreciably smaller than European breeding form *roseus*, has greyish legs with gaudy pink 'knees' and more black on the top of the bill – see fig. above.) **V**

nest on rooftop

White Stork

Black Stork

adult

juv.

juv.

Spoonbill

Flamingo

77

GREBES (order Podicipediformes, family Podicipedidae)

Recent research has shown that grebes are most closely related to flamingoes (of all!), hence their new placement here. Accomplished diving birds, but smaller than the divers and have lobed toes, short legs placed far back; tail very short. The flight is swift, with the head held low. They live on fish and aquatic insects. Build a floating nest of plant material. Clutch 2–7 eggs.

Little Grebe

Little Grebe *Tachybaptus ruficollis* L 25. Widespread and generally common on densely vegetated lakes and small rivers. A master at keeping out of sight during breeding period. Rather uniformly coloured plumage and *small size* characteristic at all times of the year. *Cheeks, chin and foreneck brownish-red* in summer plumage. *Bright whitish-yellow gape patches.* More greyish-brown in winter plumage, and pale gape not so conspicuous. White bases to secondaries ususally concealed by coverts. On breeding grounds utters loud, drawn-out, shrill cackling trills, like female Cuckoo. Contact call 'beeheeb'. **RW**

Great Crested Grebe

Great Crested Grebe *Podiceps cristatus* L 50. Locally common on inland lakes and rivers with reed cover. During migration and winter, along the coasts and on large lakes and reservoirs, usually in small flocks. *Long, thin neck, large ear tufts.* In winter plumage the ear tufts and tippets are absent, confusion can then occur with Red-necked Grebe. Distinguished from latter by *white above the eye, longer, paler neck and longer, pink bill* with dark culmen. Also, shows *much more white on its wings* in flight; frontal and rear white patches merge at the base of the wing. The courtship display is remarkable and is frequently performed. Characteristic is the breast-to-breast 'flirting' with vigorous head-shaking. Its climax is the so-called 'penguin-dance'. Calls include a far-carrying, rumbling 'korrr' (often at night) and a harsh cackling 'vrek-vrek-vrek-'. The young beg with a loud 'ping-ping-ping-'. **RW**

Red-necked Grebe

Red-necked Grebe *Podiceps grisegena* L 45. Fairly common but local on lowland lakes and shallower marshy ponds, usually with tall surrounding cover. Winters mostly on coasts, occasionally on inland lakes, reservoirs. In summer plumage unmistakable, in winter plumage easily confused with Great Crested Grebe. Distinguished by *shorter, grey neck and dark bill with yellow base.* In flight, compared with Great Crested, appears stunted at front, and the front white wing patch (quite big) does not reach inner part of rear one. Noisy in spring. The call most resembles the Water Rail's squealing call but is deeper and more intense. It starts with a Pheasant-like stutter, is then drawn out in a roaring howl. **W**

Horned Grebe

Horned Grebe (Slavonian Grebe) *Podiceps auritus* L 35. Rare breeder in Scotland, locally fairly common elsewhere in N Europe, on sheltered reedy lakes. On migration and in winter on sheltered coasts and estuaries, occasionally on inland waters. Reddish-brown neck of summer can look black at distance, but the 'shaving brushes' are always well visible. In winter plumage told from Black-necked Grebe by whiter sides of head, straight bill together with *flat crown* and *angular nape.* Most common call, heard in spring, summer and autumn, is feeble but far-carrying, plaintive, rattling 'hij-aarrr', repeated in short series. Display call trilling, but *pulsating,* each wave of whinnying sounds begins with a rapid giggling but drops in pitch and dies away nasally. **RWP**

Black-necked Grebe

Black-necked Grebe *Podiceps nigricollis* L 31. Breeds locally in N Britain (rare) on shallow, well-reeded lakes in colonies, often among Black-headed Gulls. More common on Continent. During migration and in winter on open waters, along shallow coasts and in estuaries. In summer, plumage *narrow* black *neck, high forehead* and flattened, *fan-like, slightly drooping cheek tufts* characteristic. Like Horned Grebe in winter plumage but sides of head greyer, bill slender and slightly upturned, and the forehead is steep and *the crown pointed.* Commonest call a plaintive whistle, 'ooo-eet'. **RWP**

winter

display

Little Grebe

young

Great Crested Grebe

'penguin dance'

winter

resting

young

Red-necked Grebe

winter

Horned Grebe

winter

Black-necked Grebe

winter

BIRDS OF PREY (orders Accipitriformes and Falconiformes)

Birds of prey are diurnal flesh-eaters. Most take live prey, which they catch on the ground, on the water or in the air. Many of the larger species also live on carrion. All have powerful, hooked bills (with which prey is torn into pieces) and toes with powerful sharp talons (with which prey is captured, killed and held). Sexes usually alike, but females are larger than males. In the species texts, therefore, each measurement is given as a range of variation instead of as a simple average. Individual variations occur in the colours of the plumages, in some species (mainly the buzzards) considerably so. On many occasions plumage characteristics are nevertheless the safest means of correct species identification.

All birds of prey are excellent fliers, and the larger species are often seen spiralling upwards in thermals without beating their wings. The positioning of the wings during gliding, as well as fine details of silhouette, proportions and flight, often facilitate species identifiction even at long range. The smaller species are more often seen in active flight, in which a few wingbeats are followed by gliding, but they also often soar. Certain species, particularly the Short-toed Eagle, Rough-legged Buzzard, Osprey and Kestrel, are often seen hovering (hanging in one spot in the air with fluttering wingbeats) high above the ground on the look-out for prey.

On the opposite page flight silhouettes of the different types of birds of prey that occur in Europe are shown. See also pp. 104–105 for an outline in colour of the appearance of different species in flight.

Birds of prey

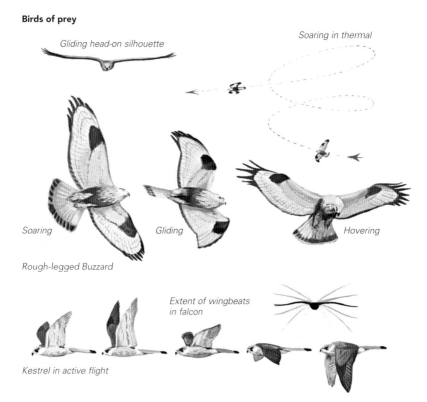

Gliding head-on silhouette

Soaring in thermal

Soaring

Gliding

Hovering

Rough-legged Buzzard

Extent of wingbeats in falcon

Kestrel in active flight

Order Accipitriformes

VULTURES are large carrion-eating birds. Two species with broad 'fingered' wings and two species with long, comparatively pointed ones. Very often seen soaring, sometimes in flocks. Nest on mountain crags or in trees. Clutches 1–2 eggs. p.82

EAGLES are large, broad-winged, broad-tailed, and for the most part, brown birds of prey. Often seen gliding. Wings usually proportionately larger than in the buzzards and with more obvious 'fingers'. The White-tailed Eagle has longer neck and shorter tail. Head and bills big. Nest in trees or on mountain crags. Clutches 1–4 eggs. p.84

BUZZARDS are medium-sized, broad-winged and broad-tailed. Plumages mainly brown, but with wide individual variations in the colours. Relatively leisurely wingbeats. Frequently soar. Often occur in flocks on migration. Nest in trees or on mountain crags. Clutches 2–6 eggs. p.90

HAWKS are medium-sized with rounded wings and long tails. Rapid wingbeats, swift active flight, but also often seen soaring. Nest in trees. Clutches 3–6 eggs. p.92

KITES are medium-sized birds of prey with long wings and long forked tails. Soar and glide skilfully, during which they frequently twist their tail as a rudder. Nest in trees. Clutches 2–4 eggs. p.94

OSPREY is large, long-winged and has very pale underparts. Often soars and hovers over water. The diet consists solely of fish. Nests in trees. Clutch 3 eggs. p.94

HARRIERS are medium-sized, long-winged and long-tailed. Often fly low over the ground with leisurely wingbeats and long glides, when the wings are held in raised position (shallow V). The plumages of the sexes differ greatly. Nest on the ground. Clutches 4 or more eggs. p.96

Order Falconiformes

FALCONS are now placed in an order of their own, and they actually appear to be, together with the parrots, closely related to the passerines. They have pointed wings and are supreme fliers, soaring a great deal. Sexes are often different. Nest on the ground, on mountain cliffs or in abandoned nests of other birds. Clutches 3–6 eggs. p.98

vulture

Golden Eagle

White-tailed Eagle

buzzard

hawk

kite

Osprey

harrier

falcon

81

Vultures

Vultures are very large and powerful birds which live principally on carrion and refuse. Their wings are very long. The vultures are for the most part seen soaring in circling flight, at times very high up, now and then making a very deep embracing wingbeat. Plumage of the sexes alike. The short-tailed species lay one egg, the wedge-tailed ones one or two. Eggs are incubated for a good seven weeks in the large species, six weeks in Egyptian Vulture.

Egyptian Vulture

Egyptian Vulture *Neophron percnopterus* L 55–65, W 155–170. Smallest of Europe's vultures (only a little bigger than Osprey). Breeds in S Europe in many habitats, most often in mountain areas. Nests in steep cliffs in crevice or small cave. Often visits refuse tips to feed. Adults easily distinguished from other raptors when seen from below on *white plumage with black flight-feathers* (though see Booted Eagle). Secondaries are partly greyish-white above. *Head small and pointed, deep yellow.* Juvenile is dark brown with attractive pattern of pale buff feather-tips. Second plumage usually more uniformly dark. White in plumage is then acquired gradually in successive moults. Dark young birds are best told on size and shape of wings (moderately broad and only slightly 'fingered') and tail (wedge-shaped). Only much larger Lammergeier has similar silhouette. Usually seen soaring, but on migration can apply persistent active flight. Roosts mostly on mountain crags. Silent. **V**

Griffon Vulture

Griffon Vulture *Gyps fulvus* L 95–110, W 230–265. Breeds locally in mountain regions in S Europe, above all in Spain, usually in small colonies (largest amounting to >100 pairs). Nests on ledges or in small caves in mountain face. *Very big*, appears to 'move in slow motion' in flight. Tail and head protrude only slightly beyond the wing edges, the *long, strongly upward-flexed 'fingers'* at the wingtips give a characteristic silhouette, especially at oblique side angles. Adults are *pale greyish-brown above*, have *whitish ruff* and *pale bill*. Young birds are rustier, have brown ruff and darkish bill. Mostly seen soaring, often several together. The wings are then *held raised above the horizontal*. Often glides away in well-spaced procession, then wings are angled at the carpal. Roosts on mountain crags, often in numbers and at certain established ledges. There, unmusical clucks and whistles can be heard. **V**

Black Vulture

Black Vulture *Aegypius monachus* L 100–115, W 245–285. Rare breeder in S Europe, mainly in Spain, in low mountains with woods and nearness of large plains. Reintroduced in S France. Nests in trees. Distinguished from the eagles by its *huge size* and the *strongly splayed 'fingers'*, from the rather similar Griffon Vulture by *slightly longer, more rounded tail* together with (and safest) *all-dark plumage* and *horizontal or even slightly down-curved wings when gliding*. Underwing-coverts almost black on immatures, becoming a shade paler with age. Rather solitary, but can be seen together with Griffons at carcases, and then occupies the highest rank in the pecking order.

Lammergeier

Lammergeier (Bearded Vulture) *Gypaetus barbatus* L 105–125, W 235–275. Very rare breeder in S Europe (< 200 pairs). Reintroduced in the Alps. Found almost exclusively in wild mountain regions. Nests in caves in inaccessible mountain faces. Easily identified *on impressive size, long and comparatively narrow wings* (tips look pointed or only moderately 'fingered') and *long, wedge-shaped tail*. From below, the contrast between pale yellowish-brown body and dark wings and tail is clearly visible. Immature is dark on head and breast, grey on the belly. Solitary in behaviour. Patrols the sides of mountains in tireless soaring flight, on the lookout for carrion. Prefers flesh from freshly killed animals. Drops bones on to a rock to break them into pieces small enough to swallow; strong gastric juices melt the bone, and the marrow is thus ingested. Usually silent, but at breeding sites gives noisy loud whistles.

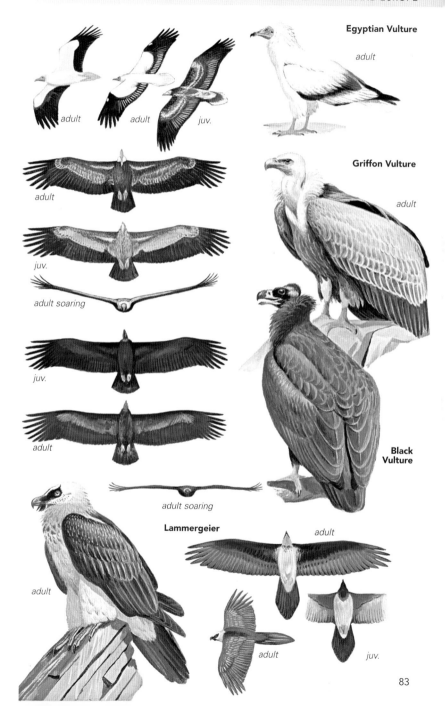

Egyptian Vulture

adult

adult

adult

juv.

Griffon Vulture

adult

adult

adult

juv.

adult soaring

juv.

adult

adult soaring

Black Vulture

Lammergeier

adult

adult

adult

adult

juv.

83

Eagles

Eagles are big, broad-winged raptors with 'fingered' wingtips and powerful bills. Often soar. The change in plumage from immature to adult takes place over several years. Sexes alike in plumage, but the female is larger.

White-tailed Eagle

White-tailed Eagle *Haliaeetus albicilla* L 77–92, W 190–240. Rare breeder in coastal regions and by lakes and rivers rich in fish. Re-introduced to W Scotland in the 1970s. The enormous stick nest is built on cliff ledges, in tall, mature pines or other large trees. A 'sluggish' eagle, spending hours perched on the look-out. In suitable weather scans while soaring, from considerable height. Lives on fish and seabirds, often carrion. Regularly robs large gulls and Ospreys. Adult has *white tail, yellow bill* and *pale brown head, neck and breast*, rest of plumage being rather dark brown. Juvenile looks generally all dark at a distance with lighter (rusty-brown) panel on median upperwing-coverts. Bill blackish with pale base (actually loral spot). Head/neck and lesser wing-coverts are dark brownish-black. The tail-feathers have dark edges but are usually pale-centred, appearing translucent (can appear quite whitish!) against the light. Also, a pale axillary patch can often be seen. Second-winter birds (c. 1½ year old) are usually much bleached: on upperwing a couple of whitish wingbars, and back and belly extensively brownish-white, blotched dark. Subsequent plumages leading to adult are generally dark brown. Much bleached immature birds may cause identification problems, but silhouette and jizz are characteristic. Often soars, on slightly arched wings. Then easily told on huge size, broad and rectangular dark wings, rather narrow and long neck, large bill and short, wedge-shaped tail (tail of juvenile clearly longer, though). More vulture-like than other European eagles. Active flight with typically long series of slow, shallow wingbeats relieved by sporadic short glides on slightly arched wings. Call 'klee klee klee klee klee…', strongly reminiscent of Black Woodpecker's spring call. **RV**

Golden Eagle

Golden Eagle *Aquila chrysaetos* L 80–93, W 190–225. Rare or scarce breeder in mountain districts, also in taiga and on sea cliffs. Nests on cliff ledges or in mature trees. Adults predominantly resident, but immatures and some older birds, especially in north and northeast, move south in winter, when found in cultivated country with scattered woodland. The second largest of Europe's eagles and by far the most powerful. Often hunts low along mountainsides or wood edges. Lives mainly of rabbits, hares, marmots, grouse and other birds, and to a considerable extent carrion. Experienced bird capable of killing a fox. Foxes walk with their tails right up in the air when they discover a Golden Eagle in the vicinity. Adult is mainly dark but has *golden nape shawl, bleached panels on upperwing-coverts* and *greyish inner part of tail*. Juvenile is dark brown with *rufous-tinged nape shawl*, pure *white wing patches* (size of which vary individually, not by age) and *white inner tail* with broad black end-band. In subsequent immature plumages white-based juvenile flight-feathers and tail-feathers are gradually replaced with grey-based adult ones, so that sub-adults have a mixture of both types. Flight powerful, usually 6–7 wingbeats followed by short glide of 1–2 seconds in regular succession (typical flight mode of all *Aquila* eagles, but different from White-tailed Eagle). In soaring flight, and frequently also when gliding, *wings are held slightly raised above horizontal* (like a very shallow V). Easily distinguished from White-tailed Eagle on silhouette by *longer and evenly rounded* (not wedge-shaped) *tail*, a little narrower and *more curved, less rectangular wings*, and by *shorter neck*. Much larger than the spotted eagles, and these soar and glide with arched and somewhat lowered wings. Steppe Eagle has shorter tail, proportionately, and has same wing posture as the spotted eagles when soaring and gliding. Main risk of confusion is with Imperial Eagle (sub-adult and adult plumages are rather similar to those of Golden), but this has slightly shorter tail, on average more evenly broad, less curved wings, and adult Imperial often soars with folded tail. **R**

nest in pine

head-on

adult

juv.

White-tailed Eagle

juv.

adult

juv.

adult

cliff nest

adult

juv.

head-on

diving

Golden Eagle

adult

juv.

adult

Imperial Eagle

Spanish
Imperial Eagle

Imperial Eagle *Aquila heliaca* L 70–82, W 175–205. Breeds in open woods within steppe areas, often at foot of mountains. Nests in trees. Adults are very *dark, blackish-brown* with *pale buffish-yellow nape shawl*, have *pure white patches on scapulars*, and *grey inner tail*. Juveniles are sandy-brown and heavily *streaked dark on breast* and mantle, contrasting to *much paler and unstreaked* lower belly and trousers, and to *very pale lower back and rump. Innermost primaries paler* than rest. Lack young Steppe Eagle's broad white wingbar below. Silhouette like Golden Eagle's but wings on average slightly more rectangular, less curved, and *tail shorter* (adult's often remarkably *closed* when soaring), and *wings usually held more horizontally* when gliding and soaring.

Spanish Imperial Eagle *Aquila adalberti* W 180–210. Closely related to Imperial Eagle restricted to SW and C Spain, endangered, population <150 pairs. Differs from Imperial Eagle on its *white leading edge to wings* in adult plumage, and on more *fox-red* and *unstreaked* juvenile plumage.

Steppe Eagle

Steppe Eagle *Aquila nipalensis* L 62–74, W 165–190. In Europe confined to Kalmuck steppe and S along Caspian Sea. Large population further east. Adults dark brown with diffuse pale nape patch and hint of pale patch at base of primaries above. *Flight-feathers usually markedly barred*, in adults with *broader trailing band*. Juveniles paler brown with dark flight-feathers, rather like juvenile Imperial Eagle, but differ on having *broad white central band along underwing*, less pale inner primaries, and more uniform brown body (lacking contrasting pale trousers and lower back of Imperial). Uppertail-coverts usually white. *Bill heavy, yellow gape prominent*. Rather long-winged with deeply 'fingered' wings. Soars and glides on slightly arched wings with drooping 'hand', like the spotted eagles.

Spotted Eagle

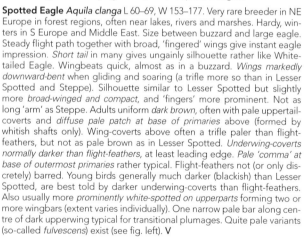

Spotted Eagle, pale
variety (v. fulvescens)

Spotted Eagle *Aquila clanga* L 60–69, W 153–177. Very rare breeder in NE Europe in forest regions, often near lakes, rivers and marshes. Hardy, winters in S Europe and Middle East. Size between buzzard and large eagle. Steady flight path together with broad, 'fingered' wings give instant eagle impression. *Short tail* in many gives ungainly silhouette rather like White-tailed Eagle. Wingbeats quick, almost as in a buzzard. *Wings markedly downward-bent* when gliding and soaring (a trifle more so than in Lesser Spotted and Steppe). Silhouette similar to Lesser Spotted but slightly more *broad-winged and compact*, and 'fingers' more prominent. Not as long 'arm' as Steppe. Adults uniform *dark brown*, often with pale uppertail-coverts and *diffuse pale patch at base of primaries* above (formed by whitish shafts only). Wing-coverts above often a trifle paler than flight-feathers, but not as pale brown as in Lesser Spotted. *Underwing-coverts normally darker than flight-feathers*, at least leading edge. *Pale 'comma' at base of outermost primaries* rather typical. Flight-feathers not (or only discretely) barred. Young birds generally much darker (blackish) than Lesser Spotted, are best told by darker underwing-coverts than flight-feathers. Also usually more *prominently white-spotted on upperparts* forming two or more wingbars (extent varies individually). One narrow pale bar along centre of dark upperwing typical for transitional plumages. Quite pale variants (so-called *fulvescens*) exist (see fig. left). **V**

Lesser Spotted Eagle

Lesser Spotted Eagle *Aquila pomarina* L 55–65, W 143–168. Breeds in healthy numbers in E Europe in woods near fields, meadows and marshes. Leaves Europe by end of Sep to spend winter in Africa. Adults dark brown with *contrasting pale yellowish-brown upperwing-coverts and head*. Also, a *prominent whitish patch at base of inner primaries above*, and pale uppertail-coverts. *Underwing-coverts clearly paler than flight-feathers* (cf. Spotted Eagle). Juveniles slightly darker, especially on head. Wing-coverts, too, are a little darker than in adults (still, generally somewhat paler than flight-feathers). Small white spots on greater-coverts form narrow, indistinct single wingbar. Small pale patch on nape can be seen at close range. Lacks the bushy 'trousers' of Spotted and Steppe).

Spanish Imperial Eagle

adult

Imperial Eagle

adult

juv.

head-on

juv.

Imperial Eagle

adult

Steppe Eagle

juv.

adult

adult

juv.

juv.

head-on

adult

Spotted Eagle

adult

juv.

adult

adult

Lesser Spotted Eagle

adult

head-on

87

Bonelli's Eagle

Bonelli's Eagle *Aquila fasciata* L 60–66, W 140–165. Scarce breeder in S Europe. Prefers open mountain regions, in winter also seen in other open terrain. Often soars in pairs along mountainsides, when generally perhaps most like Golden Eagle in behaviour and silhouette. Distinctive, however, with rather broad but comparatively not very fingered wings, with carpals projecting but rear edge fairly straight like Honey Buzzard, as well as *relatively long tail*. Wingbeats surprisingly quick (but shallow). Adult characteristic with *whitish belly* and generally *dark under-wings* (lesser underwing-coverts are whitish, bases of flight-feathers light grey, but underwings still appear dark at a distance, in contrast to belly). Tail greyish-brown with *broad dark terminal band*. Upperparts dark brown with *white patch on upper back* (like athlete's shirt number). Immature is very pale below; rosy-buff belly and coverts, greyish-white flight- and tail-feathers (closely and finely barred, no thick terminal band) with contrasting black 'fingers'. Usually there is a dark 'comma mark' around the primary coverts. In slightly older immatures there is a darker diagonal border across underwing, between coverts and flight-feathers.

Booted Eagle

Booted Eagle *Aquila pennata* L 42–49, W 110–135. Fairly common in Spain, rare elsewhere in S and central Europe. The smallest of Europe's eagles, similar in size to a Buzzard but still very much an eagle – hands markedly 'fingered'. Inhabits deciduous forest with clearings and glades, usually in lower mountain regions but also on plains. Hangs motionless in air for long periods, but does not hover. Then dives with closed wings from considerable height vertically towards ground at terrific speed (with legs extended forwards). Occurs in two colour morphs, a more common pale one and a less common dark one. Rarely, intermediate types are seen. Pale morph sometimes confused with extremely pale variants of Buzzard and Honey Buzzard but differs from these and all other raptors (except Egyptian Vulture) by underwing having *all-dark flight-feathers* behind whitish coverts. Median upperwing-coverts are, in both morphs, usually so pale that they form a *characteristically pale V on upperparts*, as in Red Kite, but in addition *uppertail-coverts are pale*. Dark morph is dark brown below (with blackish greater coverts) but with slightly paler tail, can be confused with Marsh Harrier (juv.) and Black Kite. The palest dark morph individuals (less common) are tinged red-brown, are sometimes called 'intermediate morph'. All have slightly paler, more *translucent inner three primaries*, on the whole uniform pale grey tail (darkens only slightly towards tip) and also *small white spot at front edge of each wing join* (against body sides – 'position lights'), well visible from in front. Calls shrill, clear, chattering.

Short-toed Eagle

Short-toed Eagle *Circaetus gallicus* L 66–70, W 160–180. Rather uncommon in S and E Europe but good numbers in Spain. Inhabits both mountain regions and lowland. Requires open terrain with sun-lit rocks, for lives on snakes and lizards. *A very pale, long-winged eagle*. Upperparts greyish-brown with paler wing-coverts, *underparts whitish* with small dark markings. Plumage variations: common type has quite *dark head and breast forming sharp border with whitish, narrowly cross-streaked belly*, rarer form is almost all-white below. *Never shows dark carpal patches*. The tips of the flight-feathers noticeably 'washed-out', merely grey-edged in very pale individuals (pale Buzzards have solid black tips to flight-feathers). Tail quite long and narrow with sharp corners; it has *three distinct, dark bars* (rarely a fourth is indistinctly visible at base). The wings are held horizontally in soaring flight (or very slightly raised); in gliding flight inner wing is raised and outer wing dropped with 'fingers' flexed well upwards. From below the carpal joints are seen to project pronouncedly forward. Head large, but modest in proportion to large wings. Distinguished from Osprey by broader wings and absence of dark carpal patches below. *Hovers* regularly. Majestic, thoroughly eagle-like wingbeats in normal flight. Melodic, melancholy whistle '**peeh**-o' often heard.

Bonelli's Eagle

juv.

adult

adult

adult

imm.

juv.

adult head-on

light morph

dark morph

head-on

Booted Eagle

dark morph adult

stooping

light morph adult

Short-toed Eagle

dark

light

head-on

hovering

89

Buzzards

Distinctly bigger than crows. Often seen in soaring flight. Wings broad, only moderately 'fingered'. Take small animals on the ground.

Common Buzzard

Common Buzzard *Buteo buteo* L 43–55, W 100–130. Quite common in forest and woodland, often near farmland, bogs etc. The raptor seen most often in much of N Europe. Migratory in far north. Uses fence posts and telegraph poles as look-outs, flies off with slow wingbeats and often circles in sky (*wings held raised*). Sometimes hovers. Plumage very variable. Dark forms predominate in most of Europe. In all dark plumages in Europe, characteristic *pale breast band* is apparent or at least suggested. Particularly on the Continent, white-variegated individuals exist which can be confused with Rough-legged Buzzard, Booted and Short-toed Eagles, but these pale birds often have large white wing patches above, dark comma-shaped carpal marks below. Eastern races more rusty, sometimes resemble Long-legged Buzzard. Tarsi bare. Considerable superficial resemblance to Honey Buzzard (see below). Call a mewing 'peeeeoo'. **RP**

Rough-legged Buzzard

Rough-legged Buzzard *Buteo lagopus* L 50–60, W 125–140. Fairly common breeder in northern mountains, during vole years also in adjacent forests. Winters in open flat country. *Hovers* more frequently than Buzzard, which it otherwise most resembles. Upperside typical: pale head, dark wings, gleaming *white inner part of tail*. Underparts paler, with *blackish carpal patch*. Juvenile most characteristic: much yellowish-white on breast and underwings, contrasting with *dark belly and carpal patch*. Adult males, however, can be exceedingly Buzzard-like: throat and breast are dark, belly area is usually paler, white on tail is reduced by several bars. Adult females are intermediate. Longer narrower wings and tail than Buzzard (occasionally give impression of harrier), *slower, more flexible wingbeats*, and in gliding flight the *head-on silhouette shows characteristic wing kink*. Tarsi feathered. Calls like Buzzard but more mournful. **W**

Long-legged Buzzard

Long-legged Buzzard *Buteo rufinus* L 55–62, W 130–150. Nests in SE Europe on dry steppes and in mountain districts. Large and long-winged. When gliding, wings are held slightly raised, kinked at carpal joints as in Rough-legged Buzzard. Pale head and leading edge of wing above, *pale breast, belly darkening backwards*. In adult *tail is very pale rusty-coloured without barring* (can appear white at a distance). Tail of immature pale grey-brown, barred at tip. Markings often very like eastern races of Buzzard, but difference in size and flight usually obvious. **WP**

Honey Buzzard

Honey Buzzard *Pernis apivorus* L 51–58, W 113–135. Breeds fairly commonly in woodland. Long-distance migrant. Arrives May, returns Aug/Sep. Diet first frogs, insects and young birds, later mainly larvae of wasps. Adult male above greyish-brown, *head ash-grey, eye Cuckoo-yellow*. Below usually *barred rufous* (variations: underparts all-dark brown, or almost pure white in Osprey-fashion). *Tail and flight-feathers have widely spaced bars, one at tip and two at base*. Female browner above and on head, has more dark on 'fingers' with diffuse outline, and bands on tail and wings are more evenly spaced. Juveniles even more *variable*: most are dark brown (with bright yellow cere, dark eye) – very like Common Buzzard; others are rufous-tinged or white below (then streaked on breast). Paler ones typically have *dark eye patch* and some dark on the carpals. Characteristic for juveniles to have *dark secondaries*, much dark on wingtips (more than just 'fingers' dark) and more even, narrow tail-barring. Honey Buzzard is often seen soaring, at a glance very like Common Buzzard. Silhouette, however, different: *narrower neck* (head projecting like Cuckoo's) and *longer tail* (when gliding, tightly closed, has slightly convex sides and rounded corners). Wingbeats a little deeper, more fluid and slightly slower, and *wings are held smoothly down-curved when gliding*. Call a clear, musical, melancholy '**pleee**-lu'. **SP**

display flight of Honey Buzzard

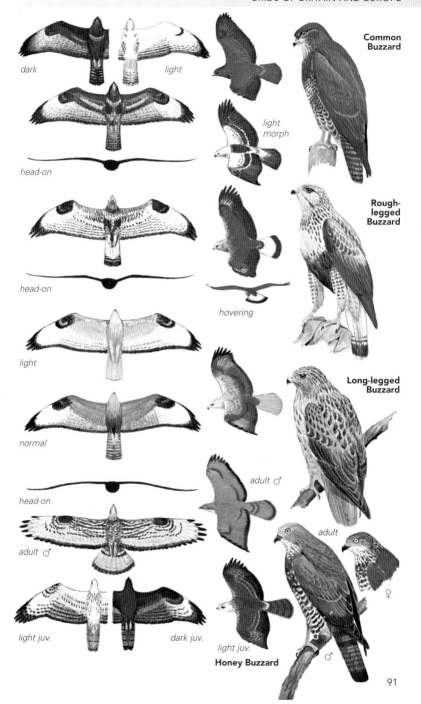

dark

light

head-on

light morph

Common Buzzard

head-on

hovering

Rough-legged Buzzard

light

normal

head-on

adult ♂

adult ♂

Long-legged Buzzard

adult

♀

light juv.

dark juv.

light juv.

Honey Buzzard

♂

Hawks

Medium-sized raptors with fairly short, rounded wings and long tails. Swift active flight, manoeuvre agilely. Gifted bird-catchers. Clutches 3–6 eggs. Nest in dense wood.

Goshawk

Goshawk *Accipiter gentilis* L 48–60, W 85–115. Breeds in woodland. Scarce in W Europe but not uncommon in N and E in mature coniferous forests. Partial migrant in far north (immatures most prone to move). Widely persecuted. Takes pigeons, crows, gamebirds, thrushes etc, often by surprise but also in very swift pursuit. Manoeuvres agilely in dense forest. Mostly, however, sweeps low over trees, crosses fields on same level in purposeful flight (rapid wingbeats, straight swift glides), often waits on concealed perch. In brighter weather often soars high up, hard to detect. From there launches impressive strike against distant quarry. Female much larger than male, has shorter wingspan than Buzzard but appears more robust. Male usually considerably larger than Carrion/Hooded Crow, in extreme cases only barely equal. Nevertheless treated with great respect by the crows: dive-bombing, but pulling out as if to avoid getting burnt, with frantic yells (not with boldness and snoring calls shown towards Sparrowhawk). Male easily confused with female Sparrowhawk. Best told by noticeably heavier, *looser wingbeats* (quick-winged in squalls and when beginning long-range attack) and *heavier body* (belly distended). Also proportionately slightly smaller head, *longer neck*, shorter tail (long enough!) with bevelled corners, longer inner wing, shorter and more pointed outer wing. *White undertail-coverts bushy*, can be spread. *Immature* heavily *streaked below* (immature Sparrowhawk barred). Display flight with slow-motion wingbeats like harrier. On perched bird profile characterised by width at 'hip' level. Alarm a loud cackling 'kyekyekye…' (Jackdaw voice!), begging call a wild, melancholy '**peee**-leh'. Both calls heard in nesting wood on early mornings in March. Mimicked by Jay. Race *buteoides* from N Russia whiter below and paler and bluer-grey above when adult; immatures have paler ground colour, show more pale mottling on upperwing. **R**

Sparrowhawk

Levant Sparrowhawk flock on migration, soaring on thermals

Sparrowhawk *Accipiter nisus* L 30–39, W 58–77. Widespread, quite common in dense woods (open country with clumps suffices). British breeders sedentary; northern ones, especially young, often move south. Takes mostly birds up to thrush size. Often flies swiftly at low level, under cover of curtains of trees etc. to make surprise attack. Drawn-out cheeps from tits often give advance warning. Quite often, however, is high up, soaring, then stoops on prey. Normal flight is *short series of quick wingbeats alternating with short descending glides*. Display with slow wingbeats (hunting flight similar at times; deceptively like Jackdaw). Female much larger than male, approaching male Goshawk, but has quick, easy wingbeats, looks much *slimmer, lighter, longer-tailed* (narrow where tail meets body). When perched shows slim body, 'padded shoulders', head drawn down. Size and dashing flight recall Merlin, size and tail length recall Kestrel, but wings rounded. Adult male slate-blue above, finely barred reddish-yellow below. Female grey above, barred grey below. Juvenile brown above, has slightly sparser, thicker barring below, upper breast almost spotted. Alarm 'kyikyikyi…', slower and jerkier than falcon. Begging call '**peee**(e)', almost identical with begging young Long-eared Owl. **RWP**

Levant Sparrowhawk

Levant Sparrowhawk *Accipiter brevipes* L 32–39, W 63–76. Inhabits open dry country in SE Europe. Feeds largely on lizards (has short, thick toes), grasshoppers etc. Tropical migrant. Migrates in *flocks* (10–30 birds, sometimes 100s). Very like Sparrowhawk. Male, however, has *sooty wingtips* contrasting with *white underparts* and lighter blue-grey upperparts (pigeon-coloured, incl. cheeks). Female also pale below with sooty wingtips. Juvenile told by *large drop-shaped spots* in vertical rows on breast. Also silhouette is slightly *shorter-tailed, with longer and more pointed wings*, thus markedly falcon-like. Iris brown.

adult soaring

♂ ♀ **Goshawk**

juv.

juv. gliding

♀

Sparrowhawk

♀

♂

soaring

juv. gliding ♂

juv. ♂ ♀

Levant Sparrowhawk

♂

soaring

juv. gliding ♂

juv. ♂ ♀

93

Kites

Kites are rather large, long-winged and long-tailed raptors that are often seen gliding. They twist their tails and manoeuvre skilfully above trees and fields. Versatile diet, including much fish, carrion and refuse. Nest in trees.

Red Kite

Red Kite *Milvus milvus* L 60–70, W 140–165. Breeds in well-wooded districts, often near lakes, fairly common in some regions in S and C Europe, also in southern-most Sweden. Recent increase in Britain. Distinguished from all other raptors by *long, deeply forked tail*, which, like the *belly, is rusty-red* (above). The wings have *large white 'windows' below*, and a broad pale brown bar across innerwing above. Young have pale streaking on breast. Distinguished from Black Kite by proportionally longer and more deeply forked tail as well as by paler and rustier plumage. Specialist in soaring and gliding flight, appears very buoyant, holds wings slightly bowed and slightly angled with projecting carpals, *turns and twists tail* continuously. In flight profile looks stooped with tail and head hanging down slightly. Call a thin piping whistle, typically undulating: 'peeeoooh, pee-oo-ee-oo-ee-oo'. **RSP**

Black Kite

Black Kite *Milvus migrans* L 50–63, W 135–150. Locally common in S and C Europe, has bred farther north. Marked preference for vicinity of wetland areas, where it snatches up fish, also in cities, for it eats refuse and carrion. Often gathers in large numbers. Resembles Red Kite in flight and proportions but less extreme. Is often easier to distinguish by tail length (shorter) than by tail fork (shallower). Plumage comparatively *uniform dark brown* including uppertail, lacking Red Kite's pale windows on underwing. The pale bar across inner upperwing is less contrasting. Immature has pale drop-shaped spotting but at a distance gives the same impression. Call resembles an immature Herring Gull's, '**pee**-errr'. **V**

Black-winged Kite *Elanus caeruleus* L 33, W 78. Breeds very scarcely and locally in dry, cultivated regions in southwestern-most Europe (recent local increases and range extension). Slightly bigger than Kestrel. Head unusually large and forward-projecting like an owl's, tail short, wings rather broad but pointed. Adult is white, pale grey and black. Immature brown-tinged on breast, neck, head and upperparts. Active flight with owl-like swift, soft wingbeats, *glides with harrier-like raised wings, often hovers.*

Black-winged Kite

Osprey (family Pandionidae)

Only a single species in this family. Widespread in many parts of the world. Threatened by environmental pollution (diet is exclusively fish) and holiday activities (nest near water).

Osprey

eagle   Osprey

nests

Osprey *Pandion haliaetus* L 53–61, W 140–165. Fairly common by freshwater lakes in N Europe and Fenno-Scandia, in Britain confined to Scottish highland lochs, in Mediterranean strictly coastal. On migration widespread inland and on coast. Builds a large stick nest on exposed site on very top of an old pine, often on a quite small island (cliffs in Mediterranean). Lives entirely on fish, for which it searches while *hovering* at height of 10–40 m and catches in a headlong dive (feet first at the strike), during which it almost completely disappears in a cascade of water. Upperparts brown with white crown, *below whitish* with a very constant wing pattern in which carpal is *always dark*. Also the secondaries appear rather dark in most lighting conditions. Female has a more obvious *breast band* than male. The Osprey is big and heavy, but the wings are neither particularly broad nor 'fingered'; at long range often gives impression of large gull, especially as wings are usually held slightly bowed. In spring the male flies on strongly undulating course with dangling feet and whistles a mournful 'yeelp-yeelp-…'. The contact call is a short loud whistle, 'pyep'. Alarm call a hoarse, sharp 'kew-kew-kew-kew'. **SP**

Red Kite

adult

juv.

juv.

head-on

adult

Black Kite

adult

juv.

juv.

adult

head-on adult

hovering

juv.

adult

Black-winged Kite

Osprey

hovering

juv.

adult

head-on

adult

95

Harriers

Medium-sized, long-winged, long-tailed raptors. Inhabit open country. Hunt low over ground (or reeds) with slow wingbeats alternating with glides on raised wings (shallow V position).

Marsh Harrier

Marsh Harrier *Circus aeruginosus* L 45–55, W 115–135. Widespread, generally uncommon but locally abundant. Nests in larger reedbeds. Hunts over swamps and cornfields. Usually seen flying low with steady wingbeats alternating with *glides on raised wings*. In early autumn juveniles soar high for hours. *Heavier and broader-winged* than other harriers, but clearly more slender than Common Buzzard. Male *pale grey on wings and tail*, may at long range gleam as a Hen Harrier, but has *rufous belly* and brown upperwing-coverts. Female dark brown with *pale yellow on head and leading edge of wing*, tail a little rufous-tinged. Juvenile darker brown with ochre cap and bib (may be missing) and all-dark wings (ochre also on leading edge of wing exceptional). Display of male is a deeply undulating dance high in the sky, with a Lapwing-like 'vay-ee' per dive. **RSP**

Hen Harrier

Hen Harrier *Circus cyaneus* L 40–50, W 100–120. Breeds on moors, in young conifer plantations, on bogs in taiga and subalpine zone. On migration and in winter, hunts over arable fields and marshes. Flight like Marsh Harrier's but also often attacks small birds in hawk-like spurts. Male characteristic (notice white rump). Female brownish with white rump, resembles Pallid and Montagu's Harriers but has *broader wings* with *four long 'fingers'* (not three), and flight is not so light. Juvenile is like female (has streaked underparts!) but colour is richer and secondaries are duskier below. Young males in moult can have wingtip resembling Pallid's (rather pointed, just a few black primaries grown in the centre). Call of male in aerial display is a chuckling series, 'chuck-uck-uck-uck'. **RWP**

Montagu's Harrier

Montagu's Harrier *Circus pygargus* L 43–48, W 102–116. Nests on lowlands moors, in swamps, cornfields and young conifer plantations. *Flight bouyant, elegant, tern-like* (especially male's). Hunting flight slow (wings raised high) in search of small rodents, lizards, insects, etc. *Wings long, slender* and *rather narrow at base, wingtip pointed* (three long 'fingers'). Sexes of similar size (though female stronger-built). Male darker grey than Pallid and Hen Harriers, has a *black bar across secondaries above* (and two below!), has *chestnut streaking on belly*. Rump greyish-white. Female brownish with white rump, readily told from Hen Harrier on shape of wings (long and pointed 'hand'!), from Pallid Harrier mainly by discernible *blackish bar across secondaries above*, a similar *dominant bar below against paler secondaries* (see fig.), *boldly chestnut-barred axillaries*, poorly developed pale neck collar and *more white above and behind eye*. Juveniles differ from Pallid Harrier on *unstreaked rufous underparts* (a few blotches on breast-sides in a few difficult to see), but is very similar to juvenile Pallid (see this). An all-dark morph appears rarely in W Europe. **SP**

Montagu's adult ♀

Pallid adult ♀

Pallid Harrier *Circus macrourus* L 39–50, W 97–120. Nests in marshes or among low bushes on steppe in SE Europe. Vagrant to N and W Europe (mostly Apr/May and Sep). Adult male characteristic: *small, pale* grey above (rump greyish), *white below* (breast grey-tinged), *black on wingtip confined to a wedge* in the centre. Female told from Hen Harrier on *narrow 'hand' with pointed tip* (three long 'fingers' instead of four), from Montagu's on *lack of bold rufous cross-barring on axillaries* (is dark brown, finely spotted paler), on *paler trailing edge to 'hand'*, and *pattern of secondaries*: dusky below (see fig.) and uniformly dark brown above. Juvenile very similar to Montagu's (unstreaked rusty below) but has more prominent and complete pale neck collar, enhanced by dark sides of neck ('boa') and on average less white above and behind eye. Second-winter male has brownish-tinged upperparts with more diffuse black wedge above. Also, silhouette and flight differ from Montagu's: slightly *shorter 'hand'*, slightly *broader 'arm'* (evenly broad to body), *quicker, stiffer, shallower wingbeats* (especially in male) and higher speed when hunting (aiming at larks, etc.), recalling light Hen Harrier with pointed wings or even a large Kestrel. **V**

Pallid Harrier

head-on

Marsh Harrier

♀

♂

juv.

♂

♀

♂

Hen Harrier

♂

Montagu's

♀

Hen
♀

1st-sum.
♂

♀

juv.

**Montagu's
Harrier**

♂

2nd-wint.

adult ♂

♂

♀

Pallid Harrier

1st-sum.

juv. Montagu's   juv. Pallid

♀

♂

♂

adult

adult

♂
2nd-
wint.

adult ♂

97

FALCONS (order Falconiformes)

Raptors with pointed wings. Flight often dashing. Female generally much larger than male.

Gyr Falcon

Gyr Falcon *Falco rusticolus* L 53–62, W 105–130. Uncommon breeder on cliff crags in mountains. Grouse are basic food, often caught in level pursuit, at times with male and female cooperating. Adults remain in the mountains all year, some immatures resort to coasts and flat country in winter. Largest and most powerful falcon. Can be confused with Goshawk (which however has shorter, more rounded, stiffer wings). Where immense mountains reduce impression of buzzard size, similarity to Peregrine is emphasised – though tail noticeably longer, *wings broader towards tips*, wingbeats slower (but can certainly move rapidly). Greyish above with *weak moustache* (more distinct in immatures) and *dirty cheeks*. Underside in adults extensively but faintly patterned, barred lower down; in immatures (which have blue-grey feet!) heavily streaked, often on yellowish-brown tinted ground colour. This applies to Scandinavian type; in Greenland a very pale, almost all-white one predominates, in Iceland it is usually between the two. A wide range of intermediates exists, however. Young in same brood can be of different colour morphs. Alarm a gruff, scolding, nasal 'geh**e**-geh**e**-geh**e**-...', more drawn-out than in Peregrine. **V**

Peregrine

Peregrine *Falco peregrinus* L 40–52, W 85–110. A widespread species which declined dramatically in second half of last century, mainly because of habitat pollution but also theft of young (for falconry) and eggs. Has now recovered markedly. Breeds mainly on cliff faces, but also on bogs and in old nests of eagles or Osprey. In winter much along coasts. Lives on medium-sized birds caught in flight. Most impressive method is the several-hundred-metre-long diagonal downward stoops with closed wings, in which the falcon appears as a blurred blob and a howling noise is created. On impact knocks out prey with its feet. Normal flight not very remarkable: quick and fairly shallow wingbeats, moderate speed. Is distinctly smaller than Gyr Falcon and Saker, and has characteristic compact silhouette with *fairly short tail* and *broad-based but sharply tapering wings*. Female larger and heavier than male; male requires more than a glance to separate from Hobby. Adult is characteristic: black and white head, *dark blue-grey upperparts*, *gleaming white breast*. Immature brown above like Lanner and Saker, but has *full moustache* and *darker crown*. Alarm a scolding 'rek-rek-rek-...'. **RP**

Lanner

Lanner *Falco biarmicus* L 40–47, W 95–110. Mainly an African species. Adapted to plains and deserts. European race *feldeggii* has declined. Majority in S Italy. Larger than Peregrine, but generally takes smaller prey. Adult *feldeggii* has *brownish-grey upperparts* with *obvious dark barring*, paler grey tail (*whole tail barred, incl. central feathers*), *thin moustache*, *rufous hind-crown* (sometimes pale!), *flanks and 'trousers' cross-barred*. Juvenile dark brown above and densely dark-streaked on underwing-coverts and breast, like Saker, but has pale '*trousers*' and *yellow feet*. Can also be confused with lighter-streaked and more compact Peregrine (especially with Arctic race *calidus*, which has thin moustache and is pale and large).

Saker

Saker *Falco cherrug* L 45–57, W 105–125. Uncommon in SE Europe, found mostly in steppe country. Lives largely on sousliks (caught by surprise during fast patrolling low over ground), also on birds. Often nests in heronries or on tall pylons. Most easily confused with Lanner, although *larger and heavier*, almost as Gyr Falcon. Adult Saker often has *much paler head* and *brown upperparts*, coverts often rufous and *paler than flight-feathers* (almost as female Kestrel). *Dark streaking* of underparts quite variable (breast often sparsely marked) but is *profuse on 'trousers'* (no cross-barring of these!). *Central tail-feathers unbarred*. Juvenile dark brown above, *heavily streaked* below, especially *on greater secondary-coverts and 'trousers'*. *Crown rather pale, dark moustache thin and often not reaching to eye*, and *feet blue-grey* as in juvenile Gyr Falcon. **V**

adult Greenland type

juv. Fenno-Scandian type

adult Fenno-Scandian race

Gyr Falcon

juv.

adult

adult

Peregrine

adult stooping

adult

juv.

adult

Lanner

adult

adult

juv.

Saker

99

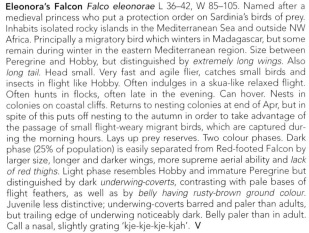

Eleonora's Falcon

Eleonora's Falcon *Falco eleonorae* L 36–42, W 85–105. Named after a medieval princess who put a protection order on Sardinia's birds of prey. Inhabits isolated rocky islands in the Mediterranean Sea and outside NW Africa. Principally a migratory bird which winters in Madagascar, but some remain during winter in the eastern Mediterranean region. Size between Peregrine and Hobby, but distinguished by *extremely long wings.* Also *long tail.* Head small. Very fast and agile flier, catches small birds and insects in flight like Hobby. Often indulges in a skua-like relaxed flight. Often hunts in flocks, often late in the evening. Can hover. Nests in colonies on coastal cliffs. Returns to nesting colonies at end of Apr, but in spite of this puts off nesting to the autumn in order to take advantage of the passage of small flight-weary migrant birds, which are captured during the morning hours. Lays up prey reserves. Two colour phases. Dark phase (25% of population) is easily separated from Red-footed Falcon by larger size, longer and darker wings, more supreme aerial ability and *lack of red thighs.* Light phase resembles Hobby and immature Peregrine but distinguished by dark *underwing-coverts,* contrasting with pale bases of flight feathers, as well as by *belly having rusty-brown ground colour.* Juvenile less distinctive; underwing-coverts barred and paler than adults, but trailing edge of underwing noticeably dark. Belly paler than in adult. Call a nasal, slightly grating 'kje-kje-kje-kjah'. **V**

Hobby

Hobby *Falco subbuteo* L 32–36, W 73–84. Breeds in fair numbers in wooded lowland with lakes, marshes and meadows. Nests in old nests of crows. Characteristic flight silhouette with long pointed wings and relatively short tail, resembles a large Swift. Extremely swift and agile hunter, often chases swallows and can even catch Swifts. Chorus of alarm is given by House Martins ('prree prree') and Swallows ('glitt, glitt'). Often dashes low over ground, when long wings are beaten in typically spaced and powerful clips. Feeds to large extent on dragonflies, which are hunted in late afternoon/evening in more relaxed flight. Never hovers. Long-distance migrant, arrives in May and disappears in Sept. Plumage characteristic with *slate-grey upperparts* and densely streaked underparts (breast and belly look uniform dark at a distance), *rusty-red thighs and undertail-coverts, white throat and cheeks, pronounced moustache.* Juvenile lacks the rusty-red, is generally more rusty-yellow in tone, often has rather pale forehead, in extreme cases so much as to create resemblance to juvenile Red-footed Falcon. Silhouette, too, is just like Red-footed's, but larger size and powerful flight often reveal the species at once. When excited, screams, long, rapid, 'over-energetic', series of notes: 'jijijijijijiji…'. Begging call '**yeee**(eh)-**yeee**(eh)…', identical with Merlin's. **SP**

Merlin

Merlin *Falco columbarius* L 25–30, W 55–65. Not uncommon breeder in subalpine birch-zone and taiga of N Europe (uses old nests of crows and raptors), rare on moors in Britain (there, nests also on ground). Smallest of Europe's raptors. Male clearly smaller than female, blue-grey above, rust-coloured below. Female is brownish above, immature too. In all plumages *diffusely marked face* with faint, indistinct moustache. Lives mainly on small birds, captured in flight after vigorous close pursuit. Often dashes low over ground, flight becoming gently bouncing (tendency towards thrush-like series of wingbeats) in final phase of attack. 'Thrush flight', however, also at low speed when new target is picked out – thus a camouflage tactic. Relatively long tail and relatively *short but pointed wings* as well as *small size* are characteristics of the flight silhouette. When the Merlin is circling very high up Peregrine is actually a perfectly possible confusion risk, since there is then nothing to compare the size with; the proportions are like Peregrine's, including obviously sturdy breast area. Alarm call fairly short, rapid, accelerating series of shrill, piercing notes (the male's more rapid, shriller). Begging call very similar to Hobby's. **RWP**

adult light morph

adult dark morph

adult light morph

juv.

Eleonora's Falcon

adult dark morph

adult

juv.

juv.

adult

Hobby

eating dragonfly

adult ♂

Merlin charging

♀

♂

Merlin

Red-footed Falcon

Red-footed Falcon *Falco vespertinus* L 28–33, W 67–76. Breeds in open country in SE Europe. Generally uncommon, but fair numbers are found on Hungarian puszta and steppes of S Russia. Nests colonially, in rook-eries. Lives mostly on insects, which are caught on ground after dive from telegraph wire, as Kestrel (also, hovers as Kestrel), or in mid-air, as Hobby (but also hunts in flocks). Often active in late evening, hunting flying cock-shafers and other insects. Flight-silhouette similar to Hobby's but smaller size and inferior speed prevents confusion. Moreover, wings are propor-tionately a little shorter, tail a little longer. Male is *slate-grey with distinctly paler flight-feathers* together with *rusty-red 'trousers'* and undertail-coverts. Female is *barred slate-grey on back* but *pale and unmarked rusty-yellow below*, and the *head is pale with contrasting dark area around eye*. Base of bill and feet are orange-red in adults. Red-footed Falcons which overshoot north and west during very hot weather in late spring are usu-ally first-summers, some already moulted to adult-like plumage (the males then slate-grey with Cuckoo-like barring on underwings, and have yellow feet), others not (the males then with juvenile head markings and grey belly with variable rusty-yellow patches). Birds of the year have dark brown back and streaked underparts (rather like Merlin) but have the same head pattern as adult female. **V**

Lesser Kestrel

Lesser Kestrel *Falco naumanni* L 28–33, W 63–72. Breeds in Spain in fair numbers, is scarce in rest of Mediterranean area and in most areas further east – has declined markedly in second half of last century. Lives in open country. Nests colonially (rarely singly or only few pairs together) in holes in buildings, both in farms and in cities, rarely in hole in ground or under boul-ders. Lives on insects. Consequently often seen dashing around in mid-air, usually in flock. Also hovers and uses telegraph-wires as look-outs. Unlike Kestrel now and then alights on the ground and walks around when feed-ing. Very like Kestrel but slightly smaller, has weaker bill and *pale claws* (Kestrel's are black). Adult male told from Kestrel by *unmarked rufous back and wings*, a *grey-blue panel on upperwing* between rufous forewing and dark flight-feathers, *lack of dark moustachial stripe*, and by practically unbarred grey-tipped flight-feathers below and pale underwing-coverts with small rounded dark spots only (can lack any spots), which make *under-wing stand out whitish against peach-coloured breast and belly*. First-summer males are more Kestrel-like, have dark-spotted rufous upperparts and lack the bluish panel on upperwing. Female has fainter moustache than Kestrel, lacks dark stripe back from eye, has thinner dark bars on rufous upperparts. Vocal at breeding sites. Quite characteristic is a three-syllable, rasping 'chay-chay-chay'. Also a Kestrel-like but faster and higher-pitched series of 'kikikikikiki'. Young beg with drawn-out trilling screams like Kestrel. **V**

Kestrel

Kestrel *Falco tinnunculus* L 32–38, W 68–78. Commonest and most wide-spread of the falcons, both in Britain and throughout Europe, familiar raptor along motorway verges. Found in almost all types of open country, from cultivated lowlands to upland moors. Frequently nests in old crow nest in a clump of trees but also on cliff ledges or even on buildings in town and city centres. Lives mainly on voles but also on insects. Scans the ground from look-out or by *hovering* at height of 7–12 m, remaining stationary in the air with fluttering wings and fanned, depressed tail. Is not particularly fast, sel-dom attempts to catch flying birds. Flight silhouette characteristic with long rather pointed wings and very long tail. Male is *rufous* (with small dark spots) *on back and wing-coverts*, has dark brown flight-feathers, *bluish-grey head and uppertail, tail with broad black terminal band*. Can be confused only with Lesser Kestrel, see that. Female and immature have rufous upperparts, and tail with prominent dark barring, and are quite difficult to separate from Lesser Kestrel in corresponding plumages. Commonest call is a piercing but not very harsh 'kikikikikiki' (in shorter series than in the other small fal-cons). Young beg with a drawn-out trilling 'keerrrl, keerrrl,…'. **RWP**

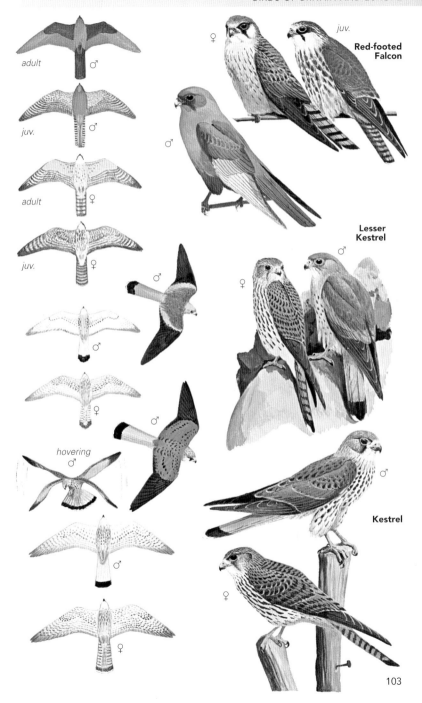

adult ♂

juv. ♂

adult ♀

juv. ♀

♂

♀

hovering ♂

♂

♂

♀

♀

♂

♂

juv.

Red-footed Falcon

Lesser Kestrel

♀ ♂

Kestrel

♂

♀

103

Birds of prey in flight

Birds of prey are often difficult to identify specifically since, because of their way of life and their shyness, they are seen at a great distance or barely for a few seconds. Details of the plumage as well as silhouette, proportions, flight and other behaviour can act as guides.

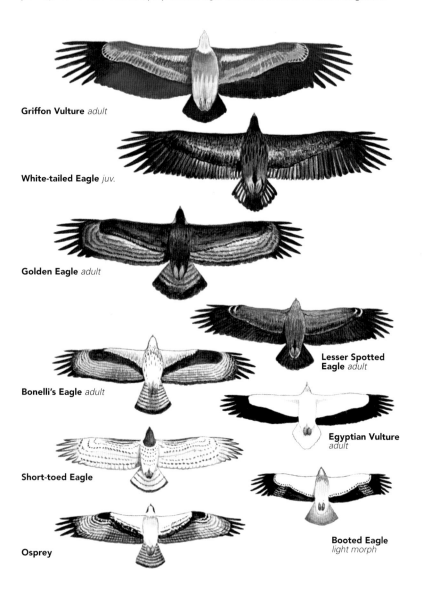

Griffon Vulture *adult*

White-tailed Eagle *juv.*

Golden Eagle *adult*

Lesser Spotted Eagle *adult*

Bonelli's Eagle *adult*

Egyptian Vulture *adult*

Short-toed Eagle

Osprey

Booted Eagle *light morph*

Since birds of prey avoid crossing open water as much as possible, during migration they often gather at headlands and along coastlines. Well-known such places are the Bosporus, Gibraltar and Falsterbo (SW corner of Sweden), where, in favourable conditions, thousands of migrating birds of prey can be seen in one day. Falsterbo is best in autumn, while the Bosporus and Gibraltar are passed by large numbers of birds of prey both in autumn and in spring. Smaller passages of certain species can be observed at various other sites in Europe, e.g. Skagen (N tip of Jutland, Denmark).

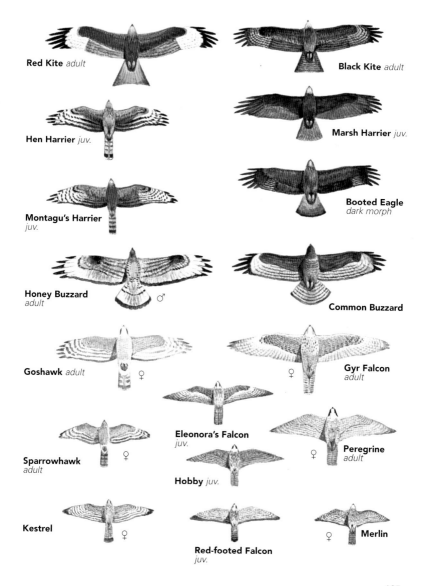

Red Kite *adult*

Black Kite *adult*

Hen Harrier *juv.*

Marsh Harrier *juv.*

Montagu's Harrier *juv.*

Booted Eagle *dark morph*

Honey Buzzard *adult* ♂

Common Buzzard

Goshawk *adult* ♀

Gyr Falcon *adult* ♀

Eleonora's Falcon *juv.*

Peregrine *adult* ♀

Sparrowhawk *adult* ♀

Hobby *juv.*

Kestrel ♀

Red-footed Falcon *juv.*

Merlin ♀

CRANES AND ALLIES

CRANES AND ALLIES (order Gruiformes)

A diverse group of birds; great variation as to size, shape and habits.

CRANES (family Gruidae) are tall, stately birds with long legs. The long neck is extended in flight. Gregarious outside the breeding season. Have a peculiar dancing display. Clutch 2 eggs. p.106

BUSTARDS (family Otididae) are large or medium-sized, long-legged and long-necked birds. They are terrestrial and prefer extensive, open plains. Gait is slow and deliberate. Very wary. Clutches 2–5 eggs. p.108

RAILS (family Rallidae) may be divided into two subfamilies: Rails and crakes are primarily wading birds, medium-sized to small, with compact bodies, long legs and long toes. Spend most of their life hidden away in tall vegetation, rather shy. Attract attention mainly by their calls. Clutches 5–15 eggs. p.110

Coots are chiefly swimming birds, duck-sized with lobed toes. Bills short and thick. Gregarious outside the breeding season. Clutches 5–12 eggs. p.112

BUTTON-QUAILS (family Turnicidae) are also members of the order Gruiformes, but they are treated on p.54 for easy comparison with similarly-looking Quail.

Crane

Crane *Grus grus* L 115–130, W 185–220. Breeds sparsely on watery taiga bogs and in reed marshes in forests, mainly in deserted regions, but sometimes unsuspectedly near farms; is unobtrusive. On migration rests on arable fields, at favoured sites in thousands. Spends the night in marshes. *Big, silvery-grey*, on the ground at great distance can look like grazing sheep; *looks 'bushy' at the rear*, the 'bush' being formed by the elongated tertials which cover the short tail. Breeding Cranes are *rusty-brown on the back* owing to their being washed with ferruginous bog water. The juvenile has a pale, rufous-tinged head without contrast (Ostrich-like) and natural brownish hue on upperparts, but has less bushy hind-part. Unlike herons, cranes *fly with extended neck*. The silver-grey can flash whitish in slanting light and recall White Stork. Migrating flocks form V or oblique line and usually fly high. Progress relaxed, includes spells of gliding (by comparison, goose Vs appear to be in a hurry). Height gained by soaring in spirals. The pair has a mating dance in spring: deep bows, high wing-flapping leaps (more peculiar than attractive). The reveille duets of the Crane pair, repeated '**krroo**krraw' and 'kaw-kaw-kaw', (extremely far-carrying, pure horn sound) are most atmospheric. In summer the breeding pair leads a secretive life, often leaves the breeding bog, guiding the young into pure wooded land. Non-breeding Cranes move south as early as late summer, breeders not until Sep. In the chorus of trumpeting, jarring 'krraw' sounds from the adults can then be heard the remarkable 'cheerp, cheerp' (like from small birds!) of the juveniles. **V**

Demoiselle Crane

Demoiselle Crane *Grus virgo* L 97–107, W 170–190. Breeds on steppe and dry high plateaux in central Asia and, sparsely west to S Russia. Is less shy than Crane. Winters, often with Cranes, on fields and in marshes, the majority in India (after spectacular passage through the passes of Hundu Kush) but also with a population in NE Africa. The latter can be seen at staging posts in Cyprus, in Aug (autumn passage thus a whole month earlier than Crane). Differs from Crane in smaller size; *long white ear tufts, elongated black breast feathers* and in *elongated tertials not being 'bushy'*. In flight, when size difficult to judge and ear tufts 'plastered down', is remarkably hard to tell from Crane, but *head looks clearly rounder*, and the black on the neck extends to below the crop. Also, wing-base proportionately a trifle broader (a little more parallel edges in Crane) and carpal area of upperwing has less contrast. Juvenile has a plain grey-white head. Calls resemble Crane's but are slightly drier and flatter. **V**

migrating

reveille

dance

juv.

Crane

adult

juv.

Demoiselle Crane

adult

Little Bustard

Little Bustard *Tetrax tetrax* L 43, W 90. Rather scarce on grassy plains or in open agricultural country in S Europe. Outside breeding season often seen in flocks, which in winter may be large. Due to moderate size not as exposed, and therefore not as habitat-demanding as Great Bustard, and as a consequence not quite as shy and unapproachable. The male discards his striking black-and-white neck pattern in winter and then resembles the female (not so different from hen Pheasant either, but belly white). Flies away with relatively *quick* (like Black Grouse) but not very propulsive *wingbeats*. Wings are hunched and rigid which adds to the impression of a gamebird. *The wings appear almost completely white* – only the four outermost primaries have much black. The 4th primary is stunted in the male and shaped in such a way that it produces a characteristic high whistling wing noise in flight. During the display in spring the cock stretches itself up with long black neck-feathers raised to a mane (cobra-look!), enhancing white pattern, and throws head back and utters the display call, a dry 'prrrt' (which sounds about as loud at 50 m as at 500 m range). It also makes fluttering leaps into the air from time to time. Display is performed mainly at dawn and dusk. **V**

Great Bustard

Great Bustard *Otis tarda* L male 100, female 80, W male 230, female 180. Inhabits large open plains (steppe or cultivated land). Has dwindled in numbers since 18th century due to hunting and habital destruction. Nowadays a fair population only in Spain, smaller ones on Hungarian puszta and on S Russian steppes, and a remnant in Germany. Formerly widespread in England, last bred in 1832 (Suffolk); attempts at re-introduction in Wiltshire currently going on. In cold winters may find its way far from its normal range. Feeds mainly on vegetable matter, but also on insects, frogs etc. Very shy and difficult to approach. The male is Europe's heaviest bird (normal weight 8–16 kg); the females are considerably lighter (3.5–5 kg). When walking on the ground the bustard may be taken at first for a roe deer or sheep. Gait deliberate. In flight looks like a giant goose with eagle wings (legs not protruding). *The wing action is powerful* and uninterrupted (no tendency to soaring flight as in the relatively light Crane). In flight *the large white wing patches* are very conspicuous. Usually occurs in flocks. On spring mornings the males display, widely dispersed over a large area – a spectacular sight. They begin by raising the tail and drawing the neck towards the back, then continue by inflating the neck like a balloon (the head is almost swamped), at the same time as displaying the long bristly moustaches straight up in the air. Large 'water lilies' of snow-white coverts under the tail and on the wings are turned forwards, and the climax is reached when the whole bird seems to be buried in a white foam bath. Several females gather around favoured males. Usually silent, but during the breeding season occasionally gives a raucous barking call. **V**

Little Bustard

display

♀

♂

display

♀

♂

♀

♂

♂

display

♀

♂

Great Bustard

Water Rail

Water Rail *Rallus aquaticus* L 28. Common in reedbeds (and in sedge bogs with tall tufts). Fairly dark in plumage, dark-spotted brown above, beautiful blue-grey below with flanks striped in black and white. Short tail, often held cocked, when shows white undertail-coverts variably tinged pale buff. *Long bill*, largely *gaudy red*. Legs long, pink-brown. Difficult to see but makes presence known by its voice. Most active at dusk and dawn. Calls are more loud than musical, usually uttered at night or when it is disturbed. In spring male utters a pounding 'kipp, kipp, kipp, kipp…' in long rhythmic series. Easily recognised is an explosive (*pig-like!*) *squealing*, which rapidly dies away, 'grruueeit, grruit, groo, gru'. Associated with this is a half-stifled groaning 'uuugh'. On spring nights a soft rumbling 'piirrr' (also the squealing call) is heard from rails flying high in the air. Female's courtship song is an associated 'bipp-bipp-biirrrr', like female Little Crake's call in structure but still distinct in its high-pitched tone and unmistakable Water Rail voice. **RSW**

Spotted Crake

Spotted Crake *Porzana porzana* L 23. Uncommon breeder on lakes with sedge and horse-tail marsh. Rare in Britain, less so on passage. Very difficult to see but may emerge on mud rim outside reeds, and then not always shy. Resembles Water Rail, both when it scampers past and when it flutters away, but has fairly short bill. Finely speckled white all over. Undertail-coverts buff. Its call carries widely on spring and summer nights: *sharp, rhythmically repeated whistles*, 'whitt, whitt, whitt, whitt…', with rhythm of just over one whistle per second. **SP**

Baillon's Crake

Baillon's Crake *Porzana pusilla* L 18. Breeds in S Europe, very rare vagrant further north. Prefers marshy meadows, swamps and pools overgrown with sedge. Secretive. Adults (sexes alike) differ from male of Little Crake in heavily barred flanks and lack of red spot at base of bill. The immature resembles immature Little Crake, but is a trifle more heavily barred and has *irregularly scattered ring-shaped white spots on the wing-coverts* (not white spots arranged in rows as in Little Crake). Also, the primary projection is very short (longer in Little Crake). *The call is low* and unmusical, *reminiscent of call of both male Garganey and several frog species*, a 2–3 second long dry rattle 'trrrrrrrrrr…' audible over 300m. **V**

Little Crake

Little Crake *Porzana parva* L 19. Breeds mainly in eastern half of Europe, rare vagrant north and west. Found in reedy swamps and ponds. Secretive. The male differs from Baillon's Crake in less strongly marked barring on the side and red spot at base of bill. Female is pale sandy-buff below, not blue-grey, and has red at base of bill. Immature resembles Baillon's – see latter for differences. The male's courtship call is a *far-carrying series of clucks*, the voice recalls barking of a small dog. Notes maintain a low tempo for a long time but then accelerate, at same time falling in pitch, and merge into a rapid stammer, 'kuak… kuak… kuak, kuak, kuak, kuak kuak-kwa-wa-a-a-a-a-a'. Female calls 'kuek-kuek-kwarrr'. **V**

Corncrake

Corncrake *Crex crex* L 26. Breeds in moist thick grass, clover and damp meadowland. Has decreased markedly over most of range. In Britain and Ireland familiar bird in 19th century, before mechanised farming, now restricted mainly to Ireland and west Scotland (but still decreasing). Suffers from grass mown early. Arrives from middle of April. Size similar to a small, slender Grey Partridge. Brownish plumage and prominent *rusty-red on the wings*, especially conspicuous in flight. Very difficult to catch sight of; runs away, concealed in the vegetation. Can be called up by imitating its voice and may then show itself briefly. If flushed, after all, the wings glisten rust-red. Flight excursions low, direct and short. Gives away its presence by its *loud, creaky and rasping, two-syllable 'rerrrp-rerrrp'* (like grating a comb on a matchbox), repeated almost once a second for hours during early summer nights (more sporadically in the daytime). **SP**

Water Rail

juv.

chick

Spotted Crake

juv.

Baillon's Crake

juv.

Little Crake

juv.

♀

♂

Corncrake

juv.

chick

Purple Swamphen *Porphyrio porphyrio* L 48. Breeds rarely in SW Europe in marshes with dense vegetation, especially larger reed-beds. Easy to identify by *chicken size, dark slaty-blue plumage, large red bill* and *long red legs*. Race in Turkey has much paler head. Immature is best identified by size and bill shape. Shy in behaviour, but more inclined to leave shelter of clumps of rushes in autumn. Very noisy, making deep tooting, nasal calls repeated in long series. Also a 'prrih prrih…'.

Purple Swamphen

Moorhen *Gallinula chloropus* L 33. Common in swamps, ponds and lakes with shore vegetation. Often seen in parks, where it walks about on grass. Places its nest in dense vegetation in or near the water. Swims readily with vigorous nodding movements. Adult easily recognised by sooty-black body colour, *red bill with yellow tip, greenish legs*, white line along the side, and *white undertail-coverts* with a black central line. The white side line and the undertail pattern also distinguish the much paler and browner juvenile from juvenile Coot. The tail is held high, and is jerked both when swimming and when walking. Short flights are low and with dangling legs; looks unstable. Nevertheless likes to make circular flights on spring nights, calling. Roosts in high bushes and trees. Downy chick black with red bill and red tonsure. Large repertoire of calls, e.g. a very characteristic sudden, bubbling 'pyurrrrk' and a sharp '**kik**ack'. May sit all night or fly around at night calling with a fast clucking 'kreck-kreck-kreck… kreck-kreck-kreck…', an alternative being a rapid, shrill, nasal 'kekeke… kekeke…'. **RW**

Moorhen

Coot *Fulica atra* L 38. Common on well-vegetated lowland lakes and ponds, nesting in reeds and other dense aquatic vegetation. In winter gregarious, found in large flocks on lakes, reservoirs and sheltered estuaries and bays. Dives frequently. Swims with nodding head movements. Runs a long distance along water surface to take off. Coots disputing territory threaten each other in hunched swimming posture, fly at each other with their feet and flap their wings so as not to fall over backwards. Adult easily recognised by *sooty-black plumage* with *white bill and frontal shield*. The immature differs from immature Moorhen in dark undertail-coverts and the characteristic Coot silhouette. Downy chick has naked red head and a ruff of yellow filaments. Many calls. Typical ones include a loud, broken 'kowk, kowk…' and an explosive 'pitts'. Coots flying around at night in spring give a nasal, trumpeting 'pe-**eh**-o'. The immatures' whimpering, lisping 'ee-lip' is distinctive. **RWP**

Coot

Red-knobbed Coot *Fulica cristata* L 40. African species. Small European population has decreased, now only < 25 pairs left. Very similar to Coot but has *two red knobs on fore-crown* at the upper edge of the white frontal shield. After breeding season the knobs shrink and are almost impossible to see at a distance. For picking out a Red-knobbed among ordinary Coots, look instead for different silhouette of swimming birds (*bulging rear part*, lower, *flattish forepart of back*, and *forward-kink of neck*), *paler legs* when walking, and *all-dark wings* when flying (Coot has whitish trailing edge). At closer range different outline of feathering at bill and shield is obvious (see fig.), and white *bill* can be seen to be *faintly tinged bluish* (pale pinkish in Coot). Resembles Coot in behaviour, but is more shy. Call disyllabic, lower-pitched than Coot's, 'keruck'.

Red-knobbed Coot

in winter:

Coot

Red-knobbed Coot

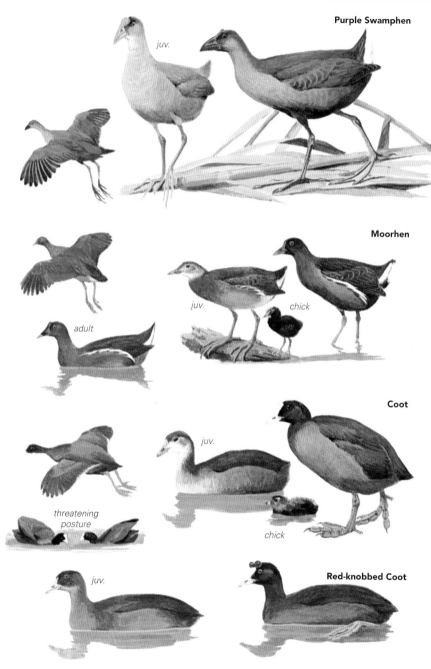

Purple Swamphen

juv.

Moorhen

juv.

chick

adult

Coot

juv.

threatening posture

chick

juv.

Red-knobbed Coot

WADERS, SKUAS, GULLS, TERNS AND AUKS

WADERS, SKUAS, GULLS, TERNS and AUKS (order Charadriiformes)

A large and varied order of birds, usually associated with water and open habitats. The group opens with the Oystercatcher (family Haematopodidae) and the lapwings (genus Vanellus).

Oystercatcher

Oystercatcher *Haematopus ostralegus* L 43. Breeds commonly along coasts, on islands and coastal meadows, locally also inland. Shorebird, which can open large bivalves with skilful incisions, but also does a lot of feeding on fields. Heavy build, unmistakably pied *black and white*, noisy. *Bill coral-red, legs pink*. A white bar across the throat in winter plumage and some juveniles. Flies low over the water, announcing itself with a shrill 'ke**beek**, ke**beek**'. The bowed wings are beaten quickly and shallowly, and flight therefore closely resembles that of duck. Migrating flocks, usually in an arc shape. Flight display has slow, stiff wingbeats like the plovers. In spring often runs around in circles in groups with open bill pointing downwards, uttering trilling 'beek, beek, beek, birrrrrrrrr-i**beek**-i**beek**…'. Alarm a shrill, short 'beek'. **RWP**

Lapwing

Lapwing *Vanellus vanellus* L 30. Breeds commonly on coastal meadows, lakeside marshy meadows, fields, arable land, moors and bogs. Breeding numbers in Britain have recently declined enormously. Easily recognised by *long thin crest*, black and white plumage (black green-glossed) and *uniquely broad, rounded wings* – males in particular have veritable 'frying-pans'! Female spotted white around bill base and on throat. Immatures have shorter crest and lack black on the throat. Early nester. Display flight of males remarkable: after a slow-motion start with deep, heaving wing-beats, the Lapwing steps up to full speed; travels along close to the ground with muffled droning wing noise ('engine throb'), fiercely pitching from side to side, suddenly to shoot up in the air with a shrill 'chay-o-wee'; then utters one or two short 'e**vip**-e**vip**' calls in normal level flight, only to dive promptly head-first towards the ground with acrobatic half-rolls and a drawn-out 'cheew-o-wee'; then continues with the frenetic zigzag flight. On breeding grounds mobs intruding humans, foxes (active at night) etc with a shrill '**weew**-ee, **weew**-ee'. Adult Lapwings migrate from the north during the summer. Migrating flocks relatively disorderly. Flight is with clipped wingbeats but leisurely for a wader. Juveniles gather in autumn in large flocks on the fields, feed on worms in plover fashion. **RWP**

Spur-winged Lapwing

Spur-winged Lapwing *Vanellus spinosus* L 28. Breeds rarely and locally in SE Europe. Frequents marshlands, sandy beaches on saltwater lakes and irrigated fields. Black and white markings make the species easy to identify. In flight distinguished from Sociable Lapwing by *black secondaries* and from White-tailed Lapwing by *black tail*. Has a hint of a crest. Behaviour like Lapwing. Wings blunt but not broad-tipped and rounded as Lapwing's. The *dark grey legs* are proportionately *longer* than in Lapwing. Alarm call is a loud, rapid 'chip-chip' and similar. Display call a fast, harsh, sonorous 'cha-radee-deeoo', superficially reminiscent of Golden Oriole's song.

Sociable Lapwing

Sociable Lapwing *Vanellus gregarius* L 30. Breeds on steppes, in serious-ly declining numbers. Very rare vagrant to W Europe. At a distance appears grey-brown, but in flight *white secondaries* and white on tail are conspicuous. Black crown, *white supercilium*, dark eye-stripe. *Belly black and chestnut-brown* in summer. Belly white in winter, then similar to Dotterel, but is bigger and has longer, blackish legs. The flight is Lapwing-like but the wing shape more normal. The calls are mostly harsh chatters, often trisyllabic, e.g. 'kretsch-kretsch-kretsch'. **V**

White-tailed Lapwing *Vanellus leucurus* L 28. Very rare vagrant from W Asia (and has bred SE Europe). Breeds at shallow, muddy lakes and calm rivers. *Amazingly long, bright yellow legs*. Plumage mainly pale sandy-brown without contrasts in face. Striking wing pattern and *all-white tail*. **V**

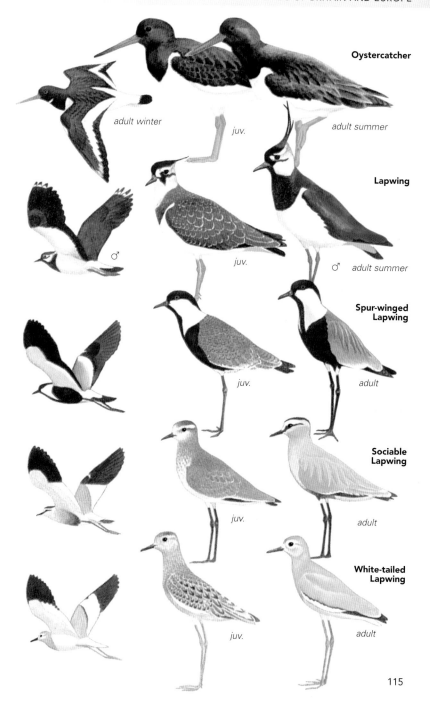

Oystercatcher

adult winter

juv.

adult summer

Lapwing

♂

juv.

♂ *adult summer*

Spur-winged Lapwing

juv.

adult

Sociable Lapwing

juv.

adult

White-tailed Lapwing

juv.

adult

Plovers (subfamily Charadriinae)

A large group of medium-sized or small waders, related to the lapwings, mainly belonging to one of two genera, Charadrius and Pluvialis. Rather stockily built with short bills. Many species have characteristic black or brown pattern on neck and breast. Long-winged, strong fliers. Nest a scrape on ground. Clutches of 3–4 eggs.

Ringed Plover

Ringed Plover *Charadrius hiaticula* L 18. Quite common, breeds along coasts on sandy and shingle beaches and short-grass shore meadow, also locally inland on heaths, by gravel-pits etc. Common on migration, alone or in small flocks, often among Dunlins. In winter common on mudflats. Appears longer-winged and flies with rather more clipped wingbeats than Dunlin. Runs very quickly ('rolling gait'). Seeks food in manner typical of small plovers (and some other plovers): alternatively trundling forward and standing dead still, suddenly bowing to pick up something. Always *white wingbar* and *pale legs* (bright orange-yellow in adults). Adult's bill orange-yellow with black tip, juvenile's dark. Call a gentle 'too-eep'. During display flight (wavering, with slow stiff wingbeats) utters a rapid murmuring 'too-widee-too-widee-…', alternating with differently stressed 'too-wedee-too-wedee- …'. RSWP

Little Ringed Plover

Little Ringed Plover *Charadrius dubius* L 16. Nests locally and sparsely on sand and gravel shores of inland waters (rarely coasts), and clay-pits and gravel-pits. Scarce even on passage. *Lacks wingbar*, always has pale legs and *dark bill*. *Yellow ring around eye* very conspicuous in the male, generally less so in the female, negligible in the juvenile. The juvenile is best distinguished from juvenile Ringed Plover by lack of white patch behind/above the eye and by yellowish-brown zone between brown crown and pale forehead (as well as by call and all-brown wings). Often markedly more active than Ringed Plover when feeding, making sallies for insects, bobbing head vigorously in between. When flying away, flight is jerkier than Ringed Plover's. Voice relatively loud and ringing. Common call 'p(i)ew' (almost monosyllabic). In wavering, slow-beating display flight (mostly at night) rapidly pounding 'pree-pree-pree-…', Sand Martin-like 'rrererere…' and drawling 'prrree-aw, prrree-aw' are heard. S

Kentish Plover

Kentish Plover *Charadrius alexandrinus* L 16. Nests on sandy shores in SE Europe, also on saline steppe lakes. Rare on passage in S and E Britain. Has proportionately large head and heavy bill. Even at long range appears *obviously pale* and white (broken breast band). Always *dark legs* typical of species (but juvenile's sometimes grey-brown). *Wingbar.* Head markings black in male, brown in female. The male's touch of orange colour on the crown and on the nape is visible at close range. Calls are a hoarse 'bee-it' (funnily like distant human wolf-whistle), 'bip, bip', which can turn into giggly 'bibibibi…', as well as rapid, hoarse 'rrererererererere…', surprisingly similar to Dunlin display call. P

Turnstone

Turnstone *Arenaria interpres* L 23. Breeds on bare rocky, stony coast, also on shore meadows with boulders, and on coastal tundra. On migration and in winter also found on sandbanks, mudflats and among seaweed, usually in small numbers. Unusual inland. Large as a Knot. Unmistakable *black, white and variegated red-brown pattern in summer plumage*. When a pair is seen at breeding site, male is distinguished by slightly brighter colours. Winter plumage is considerably duller in coloration, blackish, dark brown and white. Juvenile like adult in winter plumage but dark feathers of upperparts and breast have paler edges. Feeds at water's edge, roots about, turns over stones and seaweed by pushing with bill or head. Sits guard on elevated rocks, gives alarm with strident, shrill accelerating series of notes: 'kye-wee-kye-wee-kyewee-wetetetetet'. Other calls include short 'kew' and chuckling 'tuk-e-tuk'. WP

Ringed Plover

♀

♂

juv.

Little Ringed Plover

♂

♀

juv.

♂

♀

juv.

Kentish Plover

winter

juv.

Turnstone

summer

♂

Greater Sand Plover

Greater Sand Plover *Charadrius leschenaultii* L 23. Breeds in W and S Asia in stony deserts and on saline clayey steppes, appears in winter and on passage on sandy sea shores and saltpans. Rare vagrant in Europe. In summer plumage has *rusty-orange breast band* (of varying width; usually without black border) and black loral and forehead bands. Often also some rufous-orange on scapulars. Female usually quite different, has sandy breast band and no contrasts in face. In winter and juvenile plumages very like Lesser Sand Plover and Caspian Plover; is bigger than Lesser Sand but better distinguished from that species by *longer and more powerful bill* (proportionately bigger even than Grey Plover's), *bigger head, longer legs* (even project a little beyond tail in flight) which also are *paler, greenish-grey*. Distinguished from Caspian Plover, apart from by bill size, by white underwings, pure white outer tail feathers, *more obvious white wingbar*, incomplete greyish-brown breast patch (adult almost white in centre of breast) and also by fact that the pale supercilium is indistinct behind eye. The usual call is a fairly low, soft, rolling 'trrr', somewhat reminiscent of Turnstone. **V**

Lesser Sand Plover *Charadrius mongolus* L 20. Very rare vagrant to Europe from alpine regions in Central and E Asia. Two race groups (perhaps separate species), *mongolus* from E Siberia and *atrifrons* from Tibet, mainly differing in pattern of head and neck. Very like Greater Sand Plover but *smaller*, has *shorter, stubbier bill, shorter legs* which are darker. Moreover has *less white on tail* and *less obvious dark tail-band*. In summer plumage (sexes rather similar) has *broad rufous breast-band* and black cheeks and black forehead (*atrifrons*) or with a small white patch above bill (*mongolus*). In winter and juvenile plumages, breast is white with grey-brown sides (may recall Kentish Plover when size is not evident); face and supercilium off-white. Also resembles Caspian Plover in winter and juvenile plumages but has *white underwing*, incomplete breast-band and heavier bill. Call similar to Greater Sand Plover's, but is shorter, harder and less rolling, 'chitik' and variants. **V**

Caspian Plover *Charadrius asiaticus* L 20. Breeds on steppes, appears in winter quarters and on passage on savanna and other open grasslands. Very rare vagrant to W Europe. In manner and posture recalls Dotterel to some extent. Male in summer plumage has a broad rusty-orange breast band with thin black lower border, female has a considerably weaker-coloured and mainly greyish-brown breast band. At this time distinctions from rather similar Greater Sand Plover and Lesser Sand Plover include whiter head (*lacks black lores and bar on forehead*). In winter and juvenile plumages very like Greater and Lesser Sand Plovers, but has thin, medium-long bill, *grey underwings*, almost wholly grey tail, *faint wingbar or none at all*, almost invariably a *complete and broad, grey-brown breast-band* as well as distinct whitish supercilium, distinct also at rear of eye. Flight call 'chep', often repeated on rising, in tone rather reminiscent of Turnstone. **V**

Killdeer *Charadrius vociferus* L 25. Rare vagrant to W Europe from N America, where it is abundant and well known. Breeds on fields, preferring same habitats as the Lapwing in Europe. *Double black breast band* and size distinguish this species on the ground. In flight recognised by distinct white wingbar, *long tail and reddish-brown rump*. Call a high 'killdee' (with same tone as young Long-eared Owl). **V**

Geater Sand Plover

summer

♀ summer

juv.

♂

winter

Lesser Sand Plover

summer

winter

atrifrons

mongolus

juv.

Caspian Plover

♀ summer

juv.

summer

♂

winter

Killdeer

Grey Plover

Grey Plover *Pluvialis squatarola* L 29. Breeds on arctic tundra. Common winter visitor to coastal mudflats and sandy shores of W Europe and NW Africa. Usually seen singly or in small groups, often well spaced out (on spring migration often in big flocks). Rare inland. Adults in summer plumage have more highly variegated appearance than Golden Plover: more white on forehead and sides of breast, black of belly reaches up under wings. As in Golden Plover, female is less contrasty than male. Juvenile resembles winter adult, has grey patterning with pale yellow tinges but is not as golden-brown as juvenile Golden Plover, has *large bill*. In *flight conspicuous white tail/rump* and *pale wingbar* are distinctive, *black axillaries* unique. The call is a three-note, plaintive whistle; two variations in stress; 'pee-oo**ee**' and 'plee-**oo**ee'. **WP**

Golden Plover

Golden Plover *Pluvialis apricaria* L 27. Breeds rather commonly in upland areas on moors, heathlands, bogs and peatlands, less common in south of range. On migration and in winter mostly on ploughed fields, meadows and pasture, permanent open grassland and occasionally mudflats, usually in compact flocks (can be very large), often with Lapwings. Flies rapidly and powerfully, on migration in blunt V formation. Northern populations in summer plumage are on average more contrasting and cleaner than the southern ones. Both lack black below in winter. Juvenile is yellow-toned brown. The Golden Plover has discernible pale wingbars but *no pale colour on tail/rump. White axillaries.* Gives a *monotone* melancholy whistle at different pitches; 'peeh' and 'pluuh'. During display flight (stiff slow-beating wings) a piping, melancholy, rhythmically pumping 'plü-**eeh**-u, plü-**eeh**-u, plü-**eeh**-u…'. This is often followed by a repetitive 'per**puu**rlya-per**puu**rlya-per**puu**rlya…' (normal flight; also heard on migration). **RWP**

Pacific Golden Plover *Pluvialis fulva* L 24. Rare vagrant from Arctic Asia and NW Alaska. Visits muddy estuaries rather than fields. Very like Golden Plover, but slightly smaller, has *slimmer body, longer legs* (projecting beyond tail in flight), *longer bill*. Adult summer resembles Golden Plover but has more black below: *flanks usually barred black* on white, *sometimes all black*, and *undertail-coverts are partly black. Coarser pattern of upperparts*. Female has whitish cheeks, framed black. Juvenile, too, looks a lot like Golden. In all plumages told from Golden (but not from American Golden) on *grey axillaries* and *pale grey underwing* (not white). Difficult to tell from American Golden (see this). Main call on migration and from vagrants is whistling 'chu-it', strikingly like Spotted Redshank's, or a little softer 'tu-ip' like Ringed Plover. Another call is trisyllabic, 'tu-lee-lu'. Also has variety of more plaintive whistles recalling Golden Plover. **V**

American Golden Plover *Pluvialis dominica* L 26. Very rare vagrant from N America. Resembles Pacific Golden Plover in having *greyish underwing*, but differs in following respects: *slightly larger, longer wings* (primary tips project strikingly when standing), *shorter legs* and *shorter bill*. Adult summer has *more black on flanks* (as in Grey Plover) and undertail-coverts. Juvenile American is *greyer* than young Pacific, almost as young Grey Plover. Moreover has fairly pronounced *pale supercilium* (Dotterel-look), and projection of primaries is usually distinguishing. Calls similar to those of Pacific, and their safe separation not fully established. **V**

Dotterel

Dotterel *Charadrius morinellus* L 23. Breeds on high-lying mountain moors with lots of lichen and scree, at the upper levels for Golden Plovers. Scarce on migration (resting on fields, occasionally coasts). Female the more clearly marked and a shade larger than male, does the courting. Juveniles lack the black on belly of adults, have *lower breast pale ochrous-tinged* (adults chestnut), and have *prominent buff-white supercilium* and a *pale band across breast*. Back is blackish with rufous-ochre pattern. The incubating male is renowned for his fearlessness. Females fly around high up, calling long series of 'pwit, pwit, pwit, pwit,…'. A Dunlin-like 'keerrr' on rising. Migration call either 'pwit, pwit,…' or a disyllabic 'pee-urr'. **SP**

Grey Plover

♀

♂

juv.

juv.

Golden Plover

♀

♂

juv.

juv.

Pacific Golden Plover

juv.

juv.

American Golden Plover

juv.

juv.

♀

♂

Dotterel

juv.

juv.

Small sandpipers (family Scolopacidae)

Small in size, rather podgy, short-necked, short-legged waders. Most are Arctic breeders. Northward migration in Apr–May, southward passage mainly in Jul–Oct. Often occur in large mixed flocks (which regularly include Ringed Plovers), mostly along coasts.

Purple Sandpiper

Purple Sandpiper *Calidris maritima* L 20. Uncommon breeder on high mountain plateaus, in Arctic down to sea level. Winters locally in small flocks on rocky islets and shores. Clearly larger than Dunlin. *Darkest* of the smaller sandpipers: *uniform slate-grey* in winter plumage with *orange-yellow legs (short)* and *bill base* (conspicuous at long range); in summer plumage rather more vividly marked in dark grey and rusty-brown, with grey-brown legs. Ordinary call is a short 'kwit', often doubled. During display flight a rather Dunlin-like droning 'trrüee-trrüee-trrüee-trrüee-…', often combined with anxiety call, a shrill, hoarse, stuttering, retarding laughter, 'kewik-wik-wiwiwiwiwi-wi-wi'. **WP**

Curlew Sandpiper

Curlew Sandpiper *Calidris ferruginea* L 19. Breeds on the Siberian tundra. On southward migration fairly uncommon but regular, usually among Dunlins or Little Stints. On northward migration (from Africa) a more easterly route is chosen, is rare in W Europe in May. Adults in summer plumage unmistakable brownish-red (darker colour than Knot's), but complete summer plumage not usually seen south of the tundra; in May it is partly hidden by white fringes, and in late summer face/neck have already been moulted to grey-buff, with only breast and belly brownish-red. Juvenile (most frequently seen) resembles juvenile Dunlin but is a shade bigger, slightly longer-legged, slimmer, with a trifle longer and more decurved bill, has paler and only faintly *marked sides of neck/breast* (orange-tinted), more distinct *pale supercilium*, prominent scaling on upperparts. Pure white rump is the feature by which species is picked up in flying Dunlin flocks. Lone Curlew Sandpipers in flight often appear slightly longer-winged and with more Reeve-like clipped wingbeats than Dunlin's. Call 'kürr**it**', more twittering than Dunlin's. **P**

Dunlin

Dunlin *Calidris alpina* L 18. Uncommon and local breeder on grassy moorland with pools, lowland mosses and saltmarshes (race *schinzii*). Large flocks come from tundras in the far north-east (mainly the race *alpina*). On southward migration in large tight flocks along coastal flats, also quite common inland (muddy areas, sewage-farms, reservoirs etc). Adults in summer are *black on the belly*, in winter insipidly grey-brown above, white below, but *fairly long, slightly decurved* (outer half) *bill* is characteristic. Juvenile rather contrastingly brown-patterned above (almost like Little Stint), distinguished by *heavily grey-spotted flanks*. In Sept acquires rows of grey winter feathers interspersed in the wings. Call a wheezy whistled 'keeerrr', but call heard from feeding flocks at close range is entirely different: high 'beep-beep, beep…'. Flight display begins with gloomy 'uerrp, uerrp,…', changes to rising, strained 'rrüee-rrüee-…', ends in a falling and slightly decelerating 'rürürürürürürürü'. **RSWP**

Broad-billed Sandpiper

Broad-billed Sandpiper *Limicola falcinellus* L 16.5. Breeds sparsely on quagmires of Lapland bogs. Rare on southward as well as northward passage. Quiet and unobtrusive. Most reminiscent of a juvenile Dunlin but slightly smaller and darker. *Narrow snipe-like stripes* run down markedly *dark back*. *Dark stripes on head* even more striking; supercilium forked (in front of eye). *Bill rather long, downward-kinked at very tip*. Wingbar rather faint. Tail and rump as Dunlin. Juvenile resembles adult in summer plumage but has broader pale edges above and lacks V-shaped marks on flanks. Winter plumage like Dunlin with greyish-brown upperparts and dark carpal area but still characteristic head pattern. Call a buzzing whistle, more biting, dryer, more rasping than Dunlin's, a little like Sand Martin's: 'brrreeit'. Voice also revealed in the rhythmically buzzing display. Alternative flight call 'tett', quite distinctive. **V**

Purple Sandpiper

winter

winter

summer

juv.

Curlew Sandpiper

winter

juv.

summer

juv.

Dunlin

winter

summer

juv.

juv.

winter

summer

juv.

juv.

Broad-billed Sandpiper

Knot

Knot *Calidris canutus* L 25. Arctic species. Winters on large sandy or muddy estuaries, sometimes in huge flocks at favoured localities. On migration regular in small flocks along coast (rare inland), often with Dunlin. Much larger than Dunlin (body size as Redshank's) but typical *Calidris* in behaviour. Body almost disproportionately big and stout (long-distance flier, powerful 'engine'), *bill rather short*. Adults in summer plumage beautiful *copper-red below*. In winter plumage *pale grey* with delicate *scaly pattern* (fine black and white feather edges). Juveniles resemble adult in winter plumage, but darker grey upperparts have more distinct *scaly pattern* and breast tinged buff. *Pale grey* (vermiculated) *rump* in all plumages. Call a hoarse, nasal, squeaky 'wett-wett' (softer than Bar-tailed Godwit's). Display 'kook**l**uee, kook**l**uee,...', like subdued Curlew. **WP**

Sanderling

Sanderling *Calidris alba* L 18. Arctic species. In winter found on sandy shores, often in single-species flocks. Specialist in dashing to and fro beneath the large breakers. Most migrate through western half of Europe (May, July–Oct), stopping off at favoured localities. *Conspicuously active, constantly darting about.* Slightly bigger and stockier than Dunlin. Winter plumage extremely *pale* with *dark at bend of wing*. Juvenile also much whiter than other *Calidris*, though back vividly marked in black and rusty-yellow. Small speckled 'neck boa'. *Bill straight and relatively short. Black legs.* Appreciably *broader white wingbar* than, e.g., Dunlin. Summer plumage rusty-brown apart from white belly. In May, however, usually still has pale feather edges, giving grey and 'untidy' rather than rusty-brown impression. Call a short 'klit'. **WP**

Little Stint

Little Stint *Calidris minuta* L 14. Breeds on tundras in the north-east. Rare on northward migration in W Europe (chooses easterly route), fairly common on southward migration, on mudflats, sandy beaches, saltmarshes, also inland (muddy lakes, reservoirs etc), usually among Dunlins. Differs from latter not only in *small size* and short bill but also in more *active and scampering* behaviour. Temminck's Stint is roughly as small but behaves differently (see below). The two are also very different in appearance. Juvenile Little Stints (the ones most often seen on migration) are characteristic: have *rather a lot of white on face*, some rusty-brown on sides of breast, and *richly coloured back* (rusty-brown and black) *with two white longitudinal lines* which join to form a V. Adults are more rusty on face and breast (cf. Sanderling). When they pass south in autumn they are faded: face and breast pale yellowish-brown, on upperparts blackish centres of scapular feathers fairly striking. *Blackish legs* in all ages. Call a very thin and high 'tit'. Display (on ground or in flight) consists of weak, hissing 'svee-svee-svee- . . .', reminiscent of 'bibbling' Great Snipe. At this time also utters a silver-clear trilling 'svirrr-r-r' (actually very like Temminck's Stint). **P**

Temminck's Stint

Temminck's Stint *Calidris temminckii* L 13.5. Breeds fairly commonly to sparsely on sandy shores of lakes and rivers in northern mountain districts, mostly above tree line (a few breed N Scotland). Not uncommon on migration, seen as often in spring as in autumn. Often rests in small single-species groups and in less open sites than other *Calidris*, e.g. by small muddy pools on pasture meadows. Quiet and unobtrusive. Moves in slightly more crouched posture than Little Stint, appears to have longer body. Plumage recalls Common Sandpiper's: grey-brown upperparts, neck and upper breast. *Legs brownish-grey* (Little Stint: blackish). Adult has scattered but large black blotches on back; juveniles lack these. Unlike Little Stint, climbs high after rising, and when flushed, has more erratic flight with more clipped wingbeats. Call a high rolling 'tirrr-r-r'. In display flight male hangs still in air (5–10m) on fluttering, high-raised wings and gives 'interminable', high, rapid, twittering reeling, 'titititi…'. **P**

winter

summer

Knot

juv.

juv.

Sanderling

juv.

winter

summer

juv.

Little Stint

winter

summer

juv.

Temminck's Stint

juv.

winter

summer

juv.

Accidental small sandpipers

Least Sandpiper *Calidris minutilla* L 13. Very rare vagrant from North America. *The true dwarf of the genus.* Apart from size, also recognised by short, very *thin and all dark bill* (tip faintly down-curved), *dark crown* and *dark back* without clear scaly pattern but with faint pale lines (the upper ones sometimes forming V as in Little Stint) and *dirty-yellow to greyish-green legs*. Lores markedly dark. Supercilium faintly forked. Dark colour extends far down on forehead but does not reach bill. Poorly marked wingbar. Found in about same terrain as Temminck's Stint. Call a distinct high-pitched drawn-out 'prreep'. **V**

Long-toed Stint *Calidris subminuta* L 14. Very rare vagrant from E Asia. Small, has *pale legs*, usually olive-yellow, like Temminck's Stint and Least Sandpiper. *Tibia and toes very long*, giving moderate tempo of gait. Base of lower mandible sometimes slightly paler than rest of bill. *Brownish-grey of the forehead reaches right to the bill.* Nape rather pale, sets off the dark cap. *Vivid rusty-brown edges on crown, mantle and tertials.* Call shrill, rolling 'cherrrp'. **V**

Semipalmated Sandpiper *Calidris pusilla* L 14. Very rare vagrant from North America. Like Little Stint but a shade larger, has equally short but slightly *heavier bill* (deep base, 'blob tip' viewed head-on), has *darker lores and cheek patches*, whiter supercilium (not distinctly forked as in Little Stint). Back is not as rusty-brown in colour as in juvenile Little Stint, *lacks prominent white V marks*, is rather evenly 'scaly'. In juveniles scapulars do not have such extensively dark centres as in Little Stint, dark colour consists more of dark shaft streaks and dark crescent just before the tip (form an 'anchor'). Breast sides marked on grey-buff ground colour, occasionally the whole breast marked on warm buff ground (cf. Baird's Sandpiper). Legs blackish. Web right in between the bases of the toes (which otherwise only Western Sandpiper has). Call *short*, thin, *humming* 'chruup'. **V**

Western Sandpiper *Calidris mauri* L 15.5. Very rare vagrant to Europe from North America or E Asia. Biggest of the small 'stints', between Little Stint and Broad-billed Sandpiper. *Blackish legs.* Shares with Semipalmated feature of having *partial webbing between the toes. Long* bill, generally down-curved at tip, *separates most from Semipalmated* (but a few overlap and can be very difficult to distinguish). Rather short-winged and front-heavy. Juveniles are brightly coloured, have *rufous upper scapulars* and sometimes back, contrasting with plainer and greyer lower scapulars and wing coverts. *Pale-faced*, dark lores narrow and not prominent. In winter pale and clean grey, *streaks of head and breast sides more distinct*. Call a thin, high-pitched 'cheet', recalling White-rumped Sandpiper but shorter. **V**

White-rumped Sandpiper *Calidris fuscicollis* L 17. Rare vagrant from North America. Size and silhouette close to Baird's Sandpiper, with *shorter and straighter bill* and strikingly *long wings* (*primaries project well beyond tail*). Body elongated, *stance horizontal*. Adult summer told from Baird's on fine *dark streaks along flanks*, darker upperparts with some chestnut admixed, and *white rump*. Wingbar faint. In much paler and greyer winter plumage retains some dark flank-streaking. Flight call a startlingly thin 'tzeet', almost bat-like. **V**

Baird's Sandpiper *Calidris bairdii* L 16. Very rare vagrant from North America. Size, silhouette and stance very close to White-rumped, with *elongated body, horizontal* stance, *nearly straight bill*. Adult summer is rather *grey-brown*, resembling Temminck's Stint (and, as to colour, not silhouette, adult Semipalmated Sandpiper). Juvenile has *sandy cheeks and breast with fine dark spotting, unstreaked flanks* and prominently *scaly upperparts, feather-centres rather solidly blackish. Rump dark*, wingbar faint, legs blackish. Call a rather 'frothy' 'kreep'. **V**

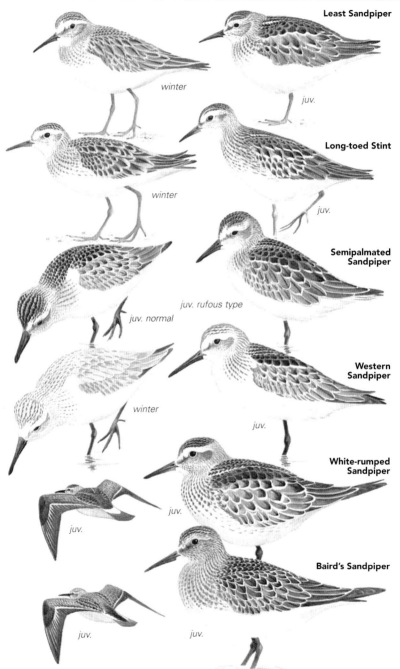

Least Sandpiper

winter

juv.

Long-toed Stint

winter

juv.

Semipalmated Sandpiper

juv. rufous type

juv. normal

Western Sandpiper

winter

juv.

White-rumped Sandpiper

juv.

juv.

Baird's Sandpiper

juv.

juv.

127

Long-billed Dowitcher *Limnodromus scolopaceus* L 30. Vagrant from E Siberia or North America. Compact build with *bill as long as Snipe* and medium-long legs. Wavy barring on rump with *white slit on back* as in Spotted Redshank. *Secondaries pale-tipped*. Very similar to Short-billed Dowitcher. Usually slightly bigger and with longer bill. In summer, whole of underparts brick-red, coarsely scalloped dark on flanks but also on sides of neck. In winter the two species can be impossible to distinguish in the field, but *dark bars on tail-feathers are usually wider than intervening white*, and *bill length* is also a supporting clue in many. Juveniles are distinguished from Short-billed Dowitchers by *cleaner grey tone on underparts* together with *uniformly dark tertials and greater-coverts with narrow pale border*. Best feature is considered to be one of the calls, a thin piping, Oystercatcher-like 'keek', uttered singly or several in a series. Vagrants seem to prefer Snipe habitats and muddy shores. **V**

Short-billed Dowitcher *Limnodromus griseus* L 29. Very rare vagrant from North America. Extremely similar to Long-billed Dowitcher. Usually slightly smaller and with shorter bill. Dark bars on tail feathers are narrower than white ones in between, or at most equal in width. In summer, plumage pale orange-red on underparts but whiter on the belly. Markings below are considerably more sparing than in Long-billed species, often only scattered spots. In winter, plumage fine differences in the tail barring and sometimes bill length are the only clues as regards appearance. In juvenile plumage it is distinguished from the Long-billed species by *warmer tone of the underparts* (buff-coloured, pale greyish-brown) together with coarsely *barred tertials* and inner greater coverts (cf. Long-billed species). The best feature is the call, a rapid slightly slurred 'tururu', somewhat recalling Turnstone (not always trisyllabic). Prefers sandy beaches but sometimes also found in Snipe habitat. **V**

Stilt Sandpiper *Calidris himantopus* L 21.5. Very rare vagrant from North America. Summer plumage characteristic with *entire underparts barred* and a *rusty-red patch behind the eye*. Winter and juvenile plumages distinguished by white rump which does *not* extend in a wedge up the back, greyish-green fairly long legs, as well as *downward-kinked tip to bill* (underparts then pale, not barred; beware of confusion with Curlew Sandpiper). Most like dowitchers in behaviour. Call a low, quite hoarse, single-note whistle. **V**

Pectoral Sandpiper *Calidris melanotos* L 19–22. Rare but regular vagrant from North America and Siberia. Clearly bigger and shorter-billed than Dunlin. Male markedly bigger than female. When alarmed, recalls a small Reeve in shape and posture (long-necked). Upperparts patterned like Little Stint. *Breast heavily streaked* on pale grey-brown ground colour, typically *sharp demarcation against white belly*. *Legs greenish- or brownish-yellow*. Bill often has paler brown base. Wings rather long, wingbar faint. Found in Snipe habitat. Wingbeats comparatively clipped. Call a rich 'drrüp', rather like Curlew Sandpiper's. **V**

Sharp-tailed Sandpiper *Calidris acuminata* L 20. Very rare vagrant from NE Siberia. Like Pectoral Sandpiper in appearance and behaviour but lacks latter's sharp border between streaked breast and white belly, has *diffuse transition*. In summer plumage has crown vividly rufous-brown streaked dark, *distinct white supercilium* and *dark line behind eye* (more contrastingly patterned head than Pectoral Sandpiper's), neck streaked, *breast and flanks richly patterned with arrowheads* on rusty-yellow ground; in winter plumage both patterning and ground colour are considerably weaker. Juvenile has only a narrow zone of streaks over lower part of neck and on breast sides on yellow-ochre ground. Bill dark. *Legs dirty-yellow*. Call soft, metallic 'weep'. **V**

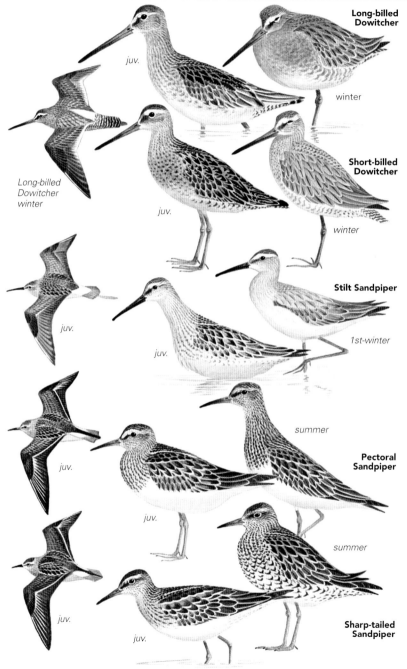

Long-billed Dowitcher

juv.

winter

Long-billed Dowitcher winter

juv.

Short-billed Dowitcher

winter

juv.

Stilt Sandpiper

juv.

1st-winter

juv.

juv.

summer

Pectoral Sandpiper

juv.

summer

juv.

juv.

Sharp-tailed Sandpiper

129

Snipes and Woodcock (family Scolopacidae)
Live in marshland and in damp wooded areas. Short legs, bills very long.

Woodcock

Woodcock *Scolopax rusticola* L 36. Common in damp woodland with open rides, also upland birchwoods in the north. On ground appears *round and short-legged* like a game bird. Rises with slight wing noise; one glimpses something *reddish-brown* among trees. Migrates at night. Male performs display flight ('roding') at dusk and dawn in spring and partly in summer. Then flies immediately above treetops with *slow wingbeats* but good speed with muffled grunting 'oo-ort, oo-ort, oo-ort', followed by explosive 'piss-p', *long bill* pointing obliquely down. Two males often hotly pursue each other giving almost twittering 'plip, plip-plip' calls. When female is put up from young, she flutters away with rear of body drooping heavily and with Jay-like scream. Can air-freight the young squeezed tight between the legs. **RSWP**

Great Snipe

Great
Snipe

young

Great Snipe *Gallinago media* L 29. Breeds uncommonly on soggy ground on mountainsides near tree line in N Europe, and at lower levels in E Europe. Has declined greatly (much hunted). Arrives May, returns Aug–Sep. Migrants use slightly drier ground than Snipe. Flushes at 4–6 m, sometimes with muffled, hoarse 'ehtch-ehtch-ehtch-…' (nothing like Snipe's call). *Flight more composed and straight* with clipped wingbeats, does not climb, drops fairly soon. Appears much *heavier* than Snipe, has much more *white on outer tail feathers*. When flying past, *more distended profile*, proportionately *slightly shorter bill, more profuse barring below* are distinctive, as well as more obvious wingbar (*white edging* also *along tips of primary coverts*, lacking in Snipe) but *hardly any white at all on trailing edge of wing*. On early summer nights displays in groups: stands erect, moves breast up and forward, opens bill wide, displays white areas of tail, all while uttering rising and falling series of rapid, high chirps ('bibbling') and a string of clicking notes (table-tennis ball!) which run into high whining 'whizzing' sounds (audible at 300 m). **V**

Snipe

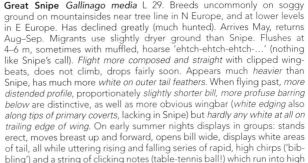

Snipe

maximum white

Snipe *Gallinago gallinago* L 25. Common in marshy areas and bogs. Most active at night. Hides in the vegetation, often in loose groups. Often rise at 10–15 m distance. As they fly up, rapidly as if catapulted up, a few scraping, explosive 'catch' notes are given. Wing action violent, flight course *pitching in zigzags. Belly and trailing edge of wing white*. Climbs to a good height, flies far away. Often sits on fence posts and calls loudly 'tik-a tik-a…' (rapid 'yikyak-yik-yak-…' notes also occur). During display flight male dives steeply with a loud humming, the so-called drumming. Sound is produced by the spread outer tail feathers, which vibrate in rapidly pulsating air current caused by wing-fluttering during the steep dive. Male also has silent display flight at lower height, in which short wing flaps are succeeded by acrobatic half-rolls. **RSWP**

Jack Snipe

Jack Snipe *Lymnocryptes minimus* L 20. Not uncommon breeder on vast Lapland bogs and locally elsewhere in far north. On passage and in winter on tussocky swamps, on flooded arable fields, etc. Extremely hard to flush, *rises at about 1m distance*. Flight then relatively fluttering, not so explosive and erratic as in Snipe. Additionally, on rising, neck more erect and tail pointed, is clearly smaller and has *considerably shorter bill*. Usually silent (but may utter a quiet 'catch'). Often drops down again quite quickly. Pronounced stripes on back but lacks pale central crown-stripe. Gives flight display in the full light of the Lapland days: moves widely around high up in shallowly undulating rising and falling flight, suddenly performs long steep dive in which a high rattling 'kollo**rap**-kollo**rap**-…' (as from a galloping horse) produced. Female has grating frog-like 'kerr'. **WP**

'roding'

Woodcock

display

Great Snipe

'drumming'

Snipe

Jack Snipe

Curlews and godwits (family Scolopacidae)
These birds occur in two distinct groups: the genera *Numenius* and *Limosa*. Large waders with long bills, which are downcurved (*Numenius*) or straight (*Limosa*).

Curlew

Curlew *Numenius arquata* L 56. Breeds fairly commonly on extensive, dry coastal dunes, lowland fields and pastures, moors and open bogs. On migration and in winter on open mudflats and shores and on coastal fields. Large as Common Gull in body and wings. Slender and tall, *long downcurved bill* (female has a noticeably longer bill than male, juvenile has relatively short bill). Drably *mottled grey-brown* with *whitish wedge-shaped rump*. Lacks eye-stripe and crown-stripes of Whimbrel, only faint suggestion is discernible. Wingbeats composed, the neck retracted, in flight at a distance recalls Common Gull. Fairly shy. Alarm call intense, rather hoarse '**kwu**wuwuwu'. Call a far-carrying, melodic, drawn-out whistle, '**kuur**-lee', on migration a more eager '**ku**ee-**ku**ee-**kuh**'. In display flight makes steep fluttering climb (silently), then glides down with 'gloomy', restrained 'oo-**ohp**, oo-**ohp**,...' notes which gradually rise in pitch and tempo and merge into clear, full, exultant, rhythmically rippling trill. **RSWP**

Whimbrel

Whimbrel *Numenius phaeopus* L 40. Breeds in mountains on cloudberry bogs and scrubby moors, mainly in far north. Passes through quickly on northward migration in Apr–May, returns July–Sep. Migrants found on mudflats, rocky shores, fields, moors (eats berries!), often with Curlews. Considerably smaller than latter, noticeably faster wingbeats, shorter bill and *brown crown with pale central stripe* together with fairly distinct dark eye-stripe. Call a shrill, *whinnying* whistle, 'puhuhuhuhuhu'. Display call has Curlew-like beginning ('oo-**ohp**') but breaks into a whinnying trill, straight and even, not pulsating: 'buurrrrrrr'. **SP**

Slender-billed Curlew *Numenius tenuirostris* L 40. On verge of extinction. Present breeding and wintering grounds unknown (historically W Siberian steppe and NW Africa, respectively), and world population now estimated to < 50. Extremely rare vagrant to S Europe. Size as Whimbrel but more *slender build*. *Bill* shape decisive: length as Whimbrel's but *narrower, tapering to fine point*. Plumage recalls Curlew's but is *paler*, especially *on secondaries and tail*. Underwing white. Adult (but not juvenile) has characteristic *heart-shaped spotting on flanks*. Call like Curlew's but higher-pitched and quicker; also, Terek Sandpiper-like trills, 'vivivivi'. **V**

Black-tailed Godwit

Black-tailed Godwit *Limosa limosa* L 40. Breeds on extensive marshes, local in distribution. On migration and in winter found along coasts and estuaries and in small numbers inland. Nervous and noisy. Flight rapid and energetic. *Long, straight bill, long legs, broad white wingbars* and *black and white tail* characteristic. Female has less rusty-red in the plumage. In winter both sexes grey-brown above, pale below. Juvenile on ground resembles juvenile Bar-tailed Godwit but neck buffier without streaking, upperparts boldly dotted rather than streaked like a Curlew. All calls are nasal, creaking and nervously repeated: 'kette**kay**', 'wiwiwi', 'weh-ee' (Lapwing-like), 'kehwee-wee**it**, kehwee-wee**it**...' (display). **RWP**

Bar-tailed Godwit

Bar-tailed Godwit *Limosa lapponica* L 38. Nests on tundra and bogs in the extreme north. Migrants found in Apr–May on coasts, and large flocks also pass through English Channel; in autumn and winter mostly small flocks seen in shallow coastal bays. Rare inland. In summer male is rusty-red on entire underparts, female is just orange-buff and distinctly larger and longer-billed. Juveniles and winter adults rather like Curlew in markings. Bill not quite so long as Black-tailed Godwit's and also *more clearly upturned, legs shorter* (esp. tibia). White wedge on back and white tail base, *tail narrowly barred dark. Lacks white wingbars*. Calls creaking, nasal. On migration 'ke-ke', at breeding site drawling or rapid series of notes: 'ku**way**-ku**way**-...' or 'ku**we**kuwe**ku**we...'. **WP**

Curlew

Whimbrel

adult

Slender-billed Curlew

adult

Black-tailed Godwit

♂ *summer*

winter

juv.

♀

♂

Bar-tailed Godwit

winter

juv.

133

Larger sandpipers (family Scolopacidae)

Medium-sized, slender waders with fairly long narrow bills and long legs. Nervous, bobbing behaviour. Often identified by their calls.

Common Sandpiper

Common Sandpiper *Actitis hypoleucos* L 20. Prefers rocky shores poor in vegetation, a common breeder by lakes, rivers and streams and on islands. Widespread on migration, both inland and on coast. Stands in characteristic, horizontal, crouched posture with *continuously rocking rear body*. Flight particularly characteristic: close above the water with *rapid, shallow wing-beats, relieved by short glides on rigid, diagonally downward-slanted wings*. Grey-brown above with white wingbars, white below with fairly pronounced grey-brown breast. Very *long tail*. Extremely reluctant to leave the vicinity of water, though often heard high over land on nocturnal migration. Passage individuals usually singly or in quite small groups. Call a thin, high 'hee-dee-dee'. Display call a series of rapid and rhythmic 'hidee**dee**deedihidi**dee**deedi-…'. Alarm 'heeep, heeep'. **SWP**

Spotted Sandpiper *Actitis macularia* L 19. Vagrant from North America. Juvenile extremely like juvenile Common Sandpiper, but has *more distinct light transverse barring on wing-coverts* (mantle on other hand less barred than Common Sandpiper), and *unmarked brown breast-sides*. In all ages shorter *wingbar, yellowish legs* and *shorter tail*. In summer plumage easily recognised by large dark spots on underparts and yellow-toned bill base. Calls quite like Common Sandpiper's but usually shorter and lower on scale and slightly rising, 'peet-weet-weeit', sometimes even resembles Green Sandpiper. Sometimes utters species-specific monosyllabic 'peet'. **V**

Green Sandpiper

Green Sandpiper *Tringa ochropus* L 23. Breeds in NE Europe, widespread but never abundant, in forests with small pools and marshes. Uses old thrush nest. Arrives early in spring, with the thaw. Females migrate south very early, in June. Migrate usually singly or 2–3 individuals together. Stops beside ponds and streams, in gullies and on watercress beds, but normally not on sedge swamps (and when it does, then e.g. by marginal pool). Resembles Wood Sandpiper but is a shade larger and broader-winged and more contrasting dark brown/white. Back *darker brown* (quite small pale spots) and *conspicuously white rump/tail*, underparts white with sharper border against speckled breast. *Brownish-black underwings*. Legs greyish-green, not quite as long as Wood Sandpiper's, feet project insignificantly beyond tail in flight, looks cut off at rear (Wood Sandpiper: comes to a point). Call a thin but clear and ringing '**tlu**eet-wit-wit'. Display call a ringing stream of shrill notes, '**tlu**ee**tu**ee-**tlu**ee**tu**ee-**tlu**ee-**tu**ee-…' or '**tee**tu**ee**-tee**tu**ee-tee**tu**ee' with introductory and interspersed alarm calls 'tit-tit-tit-tit'. **WP**

Solitary Sandpiper *Tringa solitaria* L 21. Vagrant from North America. Similar to Green Sandpiper (almost as dark underwings) but is long-winged, has *rump and central tail dark*. Also has *distinct white eye-ring*. Call a thin, often two- or three-syllable whistle, 'peet-weet-weet' a little like Common Sandpiper's. **V**

Wood Sandpiper

Wood Sandpiper *Tringa glareola* L 22. In the north quite common in sedge bogs, also in upland birch forests; in N Scotland very rare breeder near lochsides. Normally nests in sedge tussocks (exceptionally in old thrush nest). Passage Apr–May (scarce) and in Jul (fairly common), e.g. on marshy edges of lakes, usually singly but occasionally small flocks. Basically not as solitary as Green Sandpiper. Most closely resembles Green Sandpiper, but *back is not so dark brown* and is also densely and quite heavily pale-spotted, rump/tail not so pure white, breast spotting fades out on flanks, and *underwings comparatively pale. Legs olive-yellow*. The call is an excitedly repeated mellow whistle, 'chiff-chiff-chiff', display call a rapid and pleasantly ringing yodel, '**leel**tee-**leel**tee-**leel**- tee-…'. Alarm 'kip-kip-…'. **SP**

Common Sandpiper

juv.

adult

Spotted Sandpiper

juv.

adult summer

Green Sandpiper

Solitary Sandpiper

Wood Sandpiper

Greenshank

Greenshank *Tringa nebularia* L 32. Breeds fairly commonly in the north on bare, boggy moorland and open upland forest. On migration regular visitor to inland lakes and reservoirs and to shallow coastal shores and marshes (not selective), but usually only in small groups. Can be seen running after fish fry in shallow water. Our largest and most robust *Tringa. Bill relatively heavy, slightly upturned*. Neck rather pale, wings dark grey-brown, *tail/rump brilliant white*, the white continuing up the back in a broad wedge. Call is a powerful, three-note whistle, 'tew-tew-tew'. Song, given on wing and high up, a rhythmic 'clew-hü clew-hü clew-hü...'. Alarm is a fast, fierce 'kyukyukyukyu…'. **SWP**

Greater Yellowlegs *Tringa melanoleuca* L 31. Rare vagrant from North America. Recognised by large size (as Greenshank), *bright yellow legs* and white rump (*without wedge up back*). Distinguished from Lesser Yellowlegs (apart from size) by rather long, *slightly upturned bill as heavy as Greenshank's* (inner part of bill somewhat paler, like in Greenshank, especially in juvenile). Call a characteristic, sharp three- or four-syllable whistle, similar to Greenshank's but faster, more 'volatile' and often with final syllable dropping in pitch, 'chu-chu-cho'. **V**

Lesser Yellowlegs *Tringa flavipes* L 25. Rare vagrant from North America. Very like Wood Sandpiper, but slightly larger and more slender, with longer wings and longer *yellow legs*. Bill all dark. Has off-white rump (*without wedge up back*). Distinguished from Greater Yellowlegs by smaller size, shorter, *straighter and thinner bill* and also, usually, paler back. Distinguished from Marsh Sandpiper by light spots on upper parts, yellow (not greyish-green) and proportionately not quite so long legs together with lack of white wedge up back. Call a soft, very Redshank-like monosyllabic 'chu'; when flushed often a two- or three-syllable whistle. **V**

Marsh Sandpiper

Marsh Sandpiper *Tringa stagnatilis* L 23. Breeds in marshes on steppes or in taiga of E Europe. Passes through SE Europe on migration. Stops off mostly in inland areas, by pools, water meadows etc. Slightly larger than Wood Sandpiper. Slender. *Legs conspicuously long*, project a good way behind the tail in flight. *Bill markedly thin*, rather long as well as straight (very slightly upcurved in some). Markings roughly as Greenshank's: rather pale neck, brownish back/upperwing (without Wood Sandpiper's light spots, but in summer with black blotches like Reeve), white on rump and in a wedge up the back. In winter forehead white. Calls a Redshank-like 'kew', often given twice, and also a rapid series of trills, 'kewyuyuyu…' and (display) 'kuteeu-kuteeu-kuteeu-…'. **V**

Terek Sandpiper

Terek Sandpiper *Xenus cinereus* L 23. Nests along lowland rivers and lakes in Russo-Siberian taiga, up to the tundra (Kemi coast, Finland, is western outpost). Frequents shallow, muddy shores. Sporadic visitor to W Europe. Size between Common Sandpiper and Redshank. Usually stands in horizontal, crouched posture like Common Sandpiper, even rocks rear body. Quick and active when feeding, running among boulders and on driftwood, picking insects from surface. Usual flight is straight with even wingbeats, resembling Knot; at times, low over water, practices same kind of flight as Common Sandpiper. *Bill long, rather thin and noticeably upturned. Legs fairly short, yellowish or orange*. Rather a *pale grey above*, pure white below. In flight shows *broad white bar along rear edge of wing* (not so broad as Redshank's, less contrasting). *Rump and tail grey*. Common call is quick, shrill 'vivivi' with the ring of a whistle and resembling Whimbrel. Less eager variant is a subdued 'chuhuhu', like Redshank but quicker. Display call is a slow, rolling, sonorous 'klü-rrrüh, klü-rrrüh, klü-rrrüh,… ', somewhat recalling Stone-curlew display. Alarm is a smooth glissando, 'üüeet'. **V**

Greenshank

summer

juv.

Greater Yellowlegs

juv.

Lesser Yellowlegs

juv.

winter

summer

Marsh Sandpiper

juv.

summer

Terek Sandpiper

juv.

Redshank

Redshank *Tringa totanus* L 27. Typical coastal bird, commonest wader on estuaries and coastal marshes, on passage and in winter often in large flocks. Breeds in wet meadows, also inland on bogs and upland moors. Rather uniform grey-brown with *red legs* (pale orange in juvenile); cf. female Ruff and juvenile Spotted Redshank. In flight instantly recognised by *broad white bars along rear edges of wings*, conspicuous at long range also is white rump/tail. Ringing, mellow, melancholy call 'teu-hu, **teu**-huhu', song a loud '**tülle**-**tülle**-**tülle**-… chu-chu-chu… wül**yew**-wül**yew**-wül**yew**'. Alarm an irritatingly persistent 'kip-kip-kip-…', pressing on the intruder. **RSWP**

Spotted Redshank

Spotted Redshank *Tringa erythropus* L 30. Breeds chiefly in open coniferous forest in the far north. On migration and in winter found on flooded lake margins, muddy reservoirs, coastal marshes. Wades far out into the water, sometimes swims. Flight rapid with vigorous wingbeats. Summer plumage sooty-black, unmistakable (legs black, too). Immatures brown and red-legged like Redshanks, but larger, slimmer, have *longer legs* (project beyond tail, but may sometimes be held retracted) *and bill* (rather thin, straight with a *hint of a downward kink at the very tip*), are more active in behaviour. *In addition lack wingbar*, have typically narrow white 'slit' on back above a darker tail (applies to all plumages). Also more contrasting face (white line in front of eye) as well as more vermiculated flanks. Winter plumage has similar basic pattern but is grey and white instead of brown. Call a *shrill whistle*, 'chu-**it**'. On rising, occasionally gives a chuckling 'chu, chu'. Song 'trru**eeh**-e trru**eeh**-e', repeated. Alarm a rapid, dry, hard 'kekeke…', recalling angry tern. **WP**

Ruff

Ruff *Philomachus pugnax* L male 30, female 23. Breeds in sedge swamps in upland areas, most abundantly on northern tundra. On migration fairly common on marshy meadows, also on arable fields. Often tight, fairly large flocks which perform flight manoeuvres. Males (Ruffs) much larger than females (Reeves), flocks appear to consist of two different species. In May–June the males have large *ruffs and ear tufts* in different colour combinations. Gather at established sites, leks, for their remarkable display, in which at one moment they come to blows with flapping wings and the next appear to 'freeze' in deep, courtly bows. Females are light brown with big black 'diamonds' above, *legs orange-red, yellow-brown or greenish*. In late summer, young are yellowish-brown on neck, dark brown on back (pale feather-edges produce scaly pattern). Distinguished from large *Tringa* by narrow wingbars, *dark central band and white sides of rump, hunch-backed silhouette in flight* and well-spaced wingbeats, flight often includes long stretches of gliding. Often stands upright; head then looks small, and bill (slightly down-curved) short in comparison with long neck. Almost *silent* (rarely, a muffled croak). **SWP**

Buff-breasted Sandpiper *Tryngites subruficollis* L 20. Rare vagrant from North America. Resembles small Reeve; has *short bill*, small head and habit of *standing erect with extended neck*, but is *rusty-coloured buff on whole of underparts* in all plumages. *Hint of pale eye-ring*. Sides of head and neck unmarked buff. *Pale legs*. In flight, reveals *white underwing* with little dark patch on leading edge (primary coverts dark grey). Upperparts scaly like Ruff, as is rump, thus lacking white sides. No wingbar. Prefers short-grass meadows, airfields etc., but also along beaches. Call a quiet 'grreet'. **V**

Upland Sandpiper *Bartramia longicauda* L 28. Rare vagrant from North America. The size of a Redshank, brown-spotted with *very long tail* and *long, pointed, finely barred wings. Narrow neck, small head*, straight, rather short bill. Flight swift. Often alights on posts, holds wings straight up for a moment after landing. Behaviour like plovers, often rests on airfields and golf courses. Call a fast bubbling trill, 'puhuhuhuhuhuhu'. **V**

winter

Redshank

juv.

summer

juv.

winter

Spotted Redshank

juv.

summer

♀ juv.

♀
summer
('Reeve')

♂ summer

Ruff

displaying on lek

juv.

Buff-breasted

juv.

Upland

Upland Sandpiper

Buff-breasted Sandpiper

139

Avocets and stilts (family Recurvirostridae)
Elegant white and black waders with very long legs and long, thin bills. Loud calls.

Avocet

Avocet *Recurvirostra avosetta* L 43. Breeds uncommonly, though locally in fairly large, loose colonies, beside shallow sea bays, coastal lagoons and steppe lakes. *Shining white with black markings.* Slender and delicate build. Legs very long, blue-grey. *Bill thin, strongly upcurved,* is swept from side to side under the water when searching for food. Swims freely. Flies with rather fast wingbeats, not clipped, not particularly progressive. Noisy and restless. The usual call is a short, rich fluting, which is repeated with great energy, 'kluit kluit kluit…'. When young are threatened, parents give a biting and whining shriek, 'grreet', and perform injury-feigning with unusual intensity. **RSWP**

Black-winged Stilt

Black-winged Stilt *Himantopus himantopus* L 38. Nests in S Europe in shallow marshes and lagoons in small, loose colonies. Seen in small groups. Unmistakable with *improbably long, pale red legs* and *thin, straight bill,* shining white plumage with dark wings and back. Male has black back, female brown-toned. Head markings vary considerably in both sexes. Has many calls, including a persistently repeated 'krit krit krit…', recalling both Avocet and Spur-winged Plover, as well as 'krre', almost like Black-headed Gull, and Black Tern-like 'kye' or 'kyee'. **V**

Stone-curlews (family Burhinidae)
Only one species in Europe, but many in Africa. Large yellow eyes (adaptation for nocturnal habits), rather short heavy bills, long yellow legs and highly developed camouflage pattern.

Stone-curlew

Stone-curlew *Burhinus oedicnemus* L 40. Occurs sparsely in S and central Europe on dry heaths with sandy or stony areas. Now rare in Britain, confined to S and E England. Feeds on worms, insects etc, but may also take mice. Mainly nocturnal, has *large yellow eyes.* Difficult to see, runs away with head retracted and body held horizontally. Stands very erect to scan around; also tends to rest in a 'sitting' position with whole length of tarsi resting on ground but tibia vertical. When taken by surprise may flatten itself against ground with neck extended. Often seen flying away low, the size of a large Whimbrel. Flight similar to Oystercatcher's: bowed wings, beaten rather quickly, shallowly and low. *White bars and patches on the wings* show up well. At and after dusk gives its melancholy, rolling whistle, a little reminiscent of Curlew's (and Terek Sandpiper's), 'pü **pürrr**-ü **pürrr**-ü **pürrr**-ü…'. Also thin '**tü**-lee' calls like shrill Curlew, an excited 'küwü**we**-küwü**we**-küwü**we**-küwü**we**-…' like Curlew alarm call, and a 'ku**beek**-ke**beek**-ke**beek**…' like a furious outburst from an Oystercatcher. **S**

Coursers and pratincoles (family Glareolidae)
The coursers search for food by running. Only one species occurs in the area. The *pratincoles* are short-legged, have long, pointed wings, deeply forked tails and short bills. Often seen in flocks, hunt insects in flight. Clutches of 2–3 eggs.

Cream-coloured Courser *Cursorius cursor* L 23. Breeds in the Canary Islands and in the desert belt from Morocco to Pakistan. Found in the most open, most barren areas. Scans these by running with sudden halts in plover fashion (food widely scattered; much running). Usually attempts to run away from disturbance. Striking profile: narrow neck, full rear head, decurved bill which is held haughtily aloft. Sandy-coloured with black and white eye-stripes. In flight black primaries and black underwings are very striking. Wings pointed but fairly broad, flight a little Lapwing-like in looseness. Call a full and slightly nasal 'wett'. Display flight high up, uttering slowly repeated calls and the odd lower-pitched 'cheah'. **V**

Avocet

adults

juv.

Black-winged Stilt

Stone-curlew

Cream-coloured Courser

adult

juv.

adult

Collared Pratincole *Glareola pratincola* L 25. Breeds in S Europe in loose colonies in dry areas (e.g. sunbaked mud) in extensive marshy land. Spends large parts of day on the wing. Chases winged insects in elegant, fast flight (also at dawn and dusk), often many in loose party and at fairly low level. At long range appears brown with pale belly and *shining white rump/tail. Underwings reddish-brown* (also in juveniles), but this often difficult to see in the southern sun because they become shadowy black. Distinguishable from Black-winged Pratincole by narrow but distinct *white band along tips of secondaries* (can be abraded or indistinct in a few) as well as by generally slightly paler back and upperwing-coverts, which give some contrast against dark flight feathers. Can run quickly with its small, frail legs. Calls shrill, nasal, commonest flock call a five-syllable one in somewhat jerky rhythm, '**keerr**-ek-ek kit-**it**', recalls Little Tern in pitch. Also shorter conversational 'kik'. **V**

Collared Pratincole

Black-winged Pratincole *Glareola nordmanni* L 25. Very like Collared Pratincole in appearance, habits and habitat. Has, *jet-black under-wings*, generally darker upperwings, and *lacks narrow white trailing edge on inner wing*. Note that Collared too, can appear to have black underwing in strong sunlight, and that the white trailing edge on 'arm' of Collared can be abraded and thin, creating strong resemblance with Black-winged. Call like Collared Pratincole's. **V**

Black-winged Pratincole

Phalaropes (family Scolopacidae)

Small sandpipers with lobed toes, being excellent swimmers. Females have brightest colour.

Wilson's Phalarope *Phalaropus tricolor* L male 21, female 24. Rare vagrant from North America. In winter grey above, and has whitish neck and face. Juvenile dark brown above, lacking paler stripes. In all plumages *white rump* and *no wingbars*. Has *long, straight and very thin bill*, long and *strikingly thick legs*. Distinctions from Marsh Sandpiper include *lack of white wedge up back* and thicker and shorter legs. Very active. In contrast to congeners *runs* or steals, often on muddy shores and in shallow water, crouched, picking small insects from surface. **V**

Grey Phalarope *Phalaropus fulicarius* L 19. Circumpolar breeder in Arctic, also in some coastal lagoons in Iceland. Winters in South Atlantic. Sporadic visitor to coasts of W Europe, usually in autumn and winter, sometimes in hundreds after gales. Behaviour and habits like Red-necked Phalarope, fearless and almost always seen swimming. In winter plumage like Red-necked, but distinguished by *thicker and very slightly shorter bill*, which often has pale, yellowish-brown base, and by *uniform blue-grey back*. In juveniles and adults in autumn, grey back often partly variegated black, therefore resembles Red-necked more than adult in full winter plumage. Call a distinct, high 'kit', also a softer 'dreet'. **P**

Grey Phalarope

Red-necked Phalarope *Phalaropus lobatus* L 16.5. Rather common breeder in far north, in Britain rare breeder (very local) in boggy areas with small pools. Winters in large numbers in middle of Arabian Sea. Very rare on passage. Almost always seen swimming: holds neck slanting forwards like Black-headed Gull, nods in pace with swimming motions, is fussy in actions. Stirs up small animals by spinning around while swimming. Female, more showy in plumage, does the courting, leaves incubation etc to notoriously fearless male. Summer plumage unmistakable. (At distance appears dark with white chin/throat.) Juvenile white on head and neck with dark patch on crown and behind eye, and *dark on back with two pairs of rusty-yellow longitudinal stripes* (like adult in summer plumage but unlike Grey Phalarope). *Distinct white wingbar*. Winter plumage (rarely seen in Europe) like juvenile's but greyer back with white stripes. Call a short, hard 'kett' like a violin string being plucked; variations include a 'kereck', like a Coot in miniature. **SP**

Red-necked Phalarope

adult
winter

**Collared
Pratincole**

summer

**Black-winged
Pratincole**

juv.

summer

1st-winter

♀

adult winter

**Wilson's
Phalarope**

juv.

adult winter

♀

**Grey
Phalarope**

winter

juv.

Red-necked Phalarope

♀

♀ summer

juv.

winter

143

Skuas (order Charadriiformes, family Stercorariidae)

Resemble gulls but are superior in flight, feeding methods include robbing gulls and terns. In winter at sea. Central tail-feathers slightly or markedly elongated in three species.

Great Skua

Great Skua *Stercorarius skua* L 59. Breeds on islands in N Atlantic, incl. N Scotland, often in colonies on upland moors near rocky coasts. Defends nest with impressive head-on attack. Steals fish from seabirds (often Gannet) but eats carrion and offal to greater extent than do other skuas, follows fishing craft. Kills small gulls. Rather recalls dark large gull (beware confusion with oiled gull), but has much *heavier body* and *broader wing bases*, and weightiness and stability in direct flight are quite outstanding, in contrast to great agility in pursuit flight. *Large pure white wing flashes* (above as well as below) prominent even at longest ranges. **SP**

Pomarine Skua

Pomarine Skua

Pomarine Skua *Stercorarius pomarinus* L 51. Inhabits arctic tundras. In winter at sea, incl. off W Africa. Passes British coasts mainly May and Sep–Nov. In summer, adult has *long, broad tail-streamers*, which are twisted 90° so that tail tip *looks thick in side view* (but fairly thin overhead). Two morphs, one (by far commonest) pale, with or without dark breast band, one all-dark. Lack of breast band in pale morph indicates male. In winter, streamers are short but broad, not twisted, and *flanks are coarsely barred*. Juvenile very like juvenile Arctic Skua but is *larger* (= small Herring Gull; Arctic = Common Gull) and *bulkier*, with *broader wing-bases*, just exceeding tail length. Moreover, *direct flight is more powerful and steady*, recalling large gull, with *glides on more bowed wings*. Plumage usually dark, but with *more pale barring on rump* and *more white on base of under primary-coverts*. Bill is proportionally larger, and *paleness of its basal part* is often more obvious. *Central tail-feathers* are blunt and only very slightly elongated, *do not form sharp double point* (as in young Arctic). **P**

Arctic Skua

Arctic Skua

Arctic Skua *Stercorarius parasiticus*. L 46. Breeds in scattered pairs or loose colonies on barren rocky islands on Atlantic, Arctic and Baltic coasts, incl. N and W Scotland, on coastal and upland moors. Arrives in Apr, Arctic breeders pass in May, return mainly Aug–Sep. Has the *wingspan of a Common Gull*, but looks more slender. Two colour morphs, dark commonest in south, pale in north; intermediates occur. Both light morph and intermediates may have dark breast band. *Central tail-feathers elongated, thin and pointed*. Generally gull-like but always *appears strikingly dark* (even the light morph), and superior flying ability obvious. Forces gulls and terns to disgorge fish through impressive aerobatics. Even normal flight strikingly fast considering relaxed wing action. Lands on water with peculiar caution, after long glide. Juvenile varies much in colour, is very difficult to tell from Pomarine (which see), but has often a characteristic *ochrous tinge on head/neck, width of wing about equals length of tail*, or less, and *bill is more delicate and not as clearly bicoloured*. Commonest call a nasal cat-like mewing, 'eh-glaw, eh-glaw,…'. **SP**

Long-tailed Skua

Long-tailed Skua

Long-tailed Skua *Stercorarius longicaudus* L 53. Breeds on upland moor and tundra, usually far from sea. Breeding numbers vary with vole and lemming cycles (basic summer food). Parasitic habits in winter, at sea. Rare on passage (especially so in spring). In autumn, juveniles can be seen on ploughed fields, searching for earthworms. *Smallest and slimmest* of the skuas. *Wings proportionally longer and narrower, streamers thin and very long* (15–20 cm). Colour pattern of adult constant. Flight bouyant. Regularly hovers. Chases winged insects. Juvenile is utterly variable in colour, is very similar to juvenile Arctic, but tone of *colour is always cold, tinged greyish*, never ochrous-brown. Moreover size is smaller (body as Black-headed Gull) and shape more slender, with prominent breast but *long, slim rear body*. Central tail-feathers elongated but *tips blunt*. Usually white only on two outermost primary shafts. Flight weaker than Arctic's. Call in display chases a gull-like 'klee-aah', alarm a loud 'krepp-krepp-…'. **P**

juv.

adult

Great Skua

adult

tail

adult winter

adult summer

adult summer

Pomarine Skua

dark juv.

light juv.

dark

adult summer

light

tail

adult winter

adult summer

dark

light juv.

adult summer

light

Arctic Skua

dark juv.

tail

adult winter

adult summer

adult summer

light juv.

Long-tailed Skua

medium juv.

dark juv.

adult summer

Gulls (order Charadriiformes, family Laridae)

Robust birds with webbed feet, long, rather narrow wings, powerful bills, fairly short tails. Generally white, grey and black. Sexes similar. Immatures usually mottled grey-brown. Larger species gain adult plumage only after several years. Versatile, eat fish, carrion, bivalves, earthworms, birds' eggs and young etc. Nest colonially. Clutches of 2–3 eggs.

Pallas's Gull

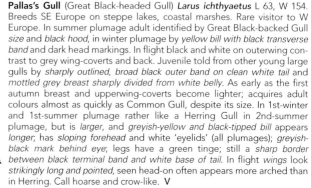

Pallas's Gull juv.

Pallas's Gull (Great Black-headed Gull) *Larus ichthyaetus* L 63, W 154. Breeds SE Europe on steppe lakes, coastal marshes. Rare visitor to W Europe. In summer plumage adult identified by Great Black-backed Gull size and *black hood*, in winter plumage by *yellow bill with black transverse band* and dark head markings. In flight black and white on outerwing contrast to grey wing-coverts and back. Juvenile told from other young large gulls by *sharply outlined, broad black outer band on clean white tail* and *mottled grey breast sharply divided from white belly*. As early as the first autumn breast and upperwing-coverts become lighter; acquires adult colours almost as quickly as Common Gull, despite its size. In 1st-winter and 1st-summer plumage rather like a Herring Gull in 2nd-summer plumage, but is *larger*, and *greyish-yellow and black-tipped bill* appears *longer*; has *sloping forehead* and white 'eyelids' (all plumages); *greyish-black mark behind eye*; legs have a green tinge; still a *sharp border between black terminal band and white base of tail*. In flight *wings* look *strikingly long and pointed*, seen head-on often appears more arched than in Herring. Call hoarse and crow-like. **V**

Glaucous Gull *Larus hyperboreus* L 61, W 150. Arctic species. In winter rare but regular on coasts, in harbours, also at refuse tips, inland reservoirs. Told from Herring Gull at all ages by *pale wingtips* and larger size (almost as big as Great Black-back). Adult is paler grey on back and wings than Herring Gull, and outermost wingtips are white. In autumn, head and nape are heavily spotted grey-brown. In 1st-winter plumage has typically *ochre-tinted ground colour*, which even at distance distinguishes it from immature Herring being mottled more grey-brown. Ground colour on breast is often so rich that a half-year-old Glaucous seen head-on appears darker than immature Herring. The spotting is slightly finer than in immature Herring, especially at very tip of tail, which is *not* darker than tail base. Immature also has *pale pink bill with black tip*. Eyes dark. Aberrant Herring Gull has slightly darker tail-tip or wingtips, dark around eye and considerably darker bill. Albinistic Herring Gulls are often all-white without either the immature's spotted patterning or the adult's grey mantle. 2nd-winter Glaucous Gull is usually *slightly* paler than juvenile plumage and lacks latter's ochre colour; in most cases recognised by more diffuse patterning, pearl-grey breaking through on mantle and wing-coverts, as well as by yellowish-brown (not dark) eye. See also Iceland Gull. Calls resemble Herring Gull's. **W**

Glaucous Gull

Iceland Gull *Larus glaucoides* L 52, W 133. Breeds in Greenland. In winter uncommon on coasts of NW Europe, in Britain also inland at refuse tips, reservoirs etc. Like Glaucous Gull in all plumages (gradual transition from pale brown to grey-white) but smaller (if anything smaller than Herring Gull), has proportionately *smaller and rounder head* and *shorter and finer bill*, therefore has a 'nicer', not so 'mean' look, more like Common Gull. Also, has *shorter legs* and *longer wings*, tips of which often project far beyond tail (Glaucous Gull's usually project less beyond). Often holds wings slightly drooping. *Bill in 1st-winter plumage darker than Glaucous Gull's*: more black at tip, often blends into paler base. In 2nd-winter plumage, bill pattern identical with Glaucous. Looks rather broad-winged and short-necked in flight. For confusion with miscoloured Herring Gulls, see Glaucous Gull. **W**

Iceland Gull

Pallas's Gull

adult summer

1st-winter

2nd-winter

adult summer

2nd-winter

1st-winter

Glaucous Gull

adult summer

1st-winter

adult

2nd-winter

1st-winter

Iceland Gull

1st-winter

adult

2nd-winter

adult summer

1st-winter

Great Black-backed Gull

Great Black-backed Gull *Larus marinus* L 63, W 155. Widespread breeder along NW European coasts, also sporadically at larger inland waters, in isolated pairs or small colonies, often with other gulls. Distinguished from rather similar Lesser Black-backed Gull by *larger size*, more *powerful, deep bill, pinkish-grey legs* and appreciably *broader wings*. Adults have *more white on wingtips*, and in W Europe have darker upperwing and back than Lesser Black-backs breeding in the same area, whereas in the Baltic it is the other way around. In winter mainly white-headed. Juvenile resembles juvenile Herring Gull (pale-patterned greater coverts, hint of pale inner primaries), but larger size and heavier bill are generally obvious, trailing *tail-band is narrower* and more broken, and head and neck is slightly paler on average (juvenile Herring is often greyer with more 'blurred' pattern). In autumn Great Black-backs (= first-winter plumage), *head and neck* become *even more whitish, pattern of mantle coarser and more contrasting*, and *tertials dark with broad whitish tips*. Confusion with smaller and darker juvenile Lesser Black-backed Gull is less likely (cf that species). Black back acquired in second winter, black upperwing in third. Flight clearly heavier than in Herring Gull, with slower wingbeats. Feeds on fish, offal and eggs and young of birds, can also kill full-grown medium-sized ducks. Calls are hoarse, gruff and very deep but not as loud as Herring Gull's. Courtship call is a slower, shorter and deeper series than in Herring Gull, lacking the initial wailing note. **RSW**

Lesser Black-backed Gull

Lesser Black-backed Gull *Larus fuscus* L 52, W 130. Breeds commonly on N and NW European coasts, locally also at inland waters. Three races: British *graellsii* with *slate-grey back and upperwing* (somewhat larger), Baltic *fuscus* with *jet-black* back and upperwing (somewhat smaller) and W Scandinavian *intermedius* which is an intermediate. Race *graellsii* is partly resident, partly migratory, *fuscus* is a long-distance migrant, crossing Europe to winter in E Mediterranean and E Africa. Adults broadly resemble adult Great Black-backed Gull (although *graellsii* has much greyer back, *fuscus* clearly blacker back), but in flight, *narrower and more pointed wings* are striking, apart from smaller size, and note black wingtips with *only one small white spot*. When standing, note *yellow legs* (not pinkish-grey), *elongated general shape* (has far-projecting wingtips), shorter legs compared with Great Black-back, *smaller head and bill*, and wingtips looking all black. In autumn, adults attain dark streaking on head and neck. Juvenile resembles juvenile Herring Gull but has darker scapulars and upperwings; *flight-feathers are uniformly all dark, lacking a pale 'window' at inner primaries*, also *outer greater coverts are much darker*. Further, underwings are darker, and the *tail-band is broad and solidly black*, contrasting well to whitish inner-tail. Dark back progressively acquired from first summer. Calls are a little deeper in pitch and more nasal in tone than Herring Gull's. **RSWP**

Heuglin's Gull

Heuglin's Gull *Larus (fuscus) heuglini* L 54, W 133. Breeds mainly on inland tundra in far NE Europe (extreme N Russia, E Kola, N White Sea and eastwards), wintering in Middle East, Arabia, W India. Migration mainly follows Russian rivers, but a few birds visit Finland and the Baltic also. Closely related to Lesser Black-backed Gull and generally treated as a subspecies of it. Extremely similar to British race *graellsii* of Lesser Black-back, with similar *slate-coloured upperparts* in adult plumage (some are a fraction paler). On average subtly larger size, longer legs and wings, and slightly longer and stronger bill than *graellsii*, but much overlap and safe separation of adults rarely possible. First winter resembles both Caspian Gull and Lesser Black-back in same plumage, sharing with these white rump, dark tail, dark tertials and dark flight-feathers. Generally differs from Lesser Black-back on being paler with rather light head/neck and on boldly patterned rather than all-dark inner greater coverts and scapulars (in these respects more like Herring); however, there is some variation and overlap, making safe identification problematic.

adult

Great Black-backed Gull

adult

1st-winter

1st-winter

2nd-summer

adult W Scandinavian race

Lesser Black-backed Gull

adult Baltic race

adult British race

juv.

juv.

Heuglin's Gull

1st-winter

adult

adult

149

Herring Gull

Herring Gull *Larus argentatus* L 58, W 140. Breeds commonly in colonies or in isolated pairs along sea coasts but also by inland lakes. Adult distinguished from similarly plumaged Common Gull by larger size, broader wings, *slower*, *'lazier' wingbeats*, *yellow eyes* and *bright yellow bill with red spot near tip*. Legs are greyish-pink. In autumn, head is streaked brown-grey. Juveniles differ from juvenile Lesser Black-backed Gull on paler wings, especially *paler inner primaries, creating a 'window'*, from juvenile Great Black-backed Gull on smaller size, less deep bill, slightly darker and more 'blurred' brown-grey plumage pattern and broader black tail-band with less contrast to pale rump. Further, duskier head is an obvious difference in first autumn, when Great Black-back has attained a whitish head. Pearl-grey back is acquired progressively from second autumn. Abundant guest in harbours and at rubbish tips, especially in winter. Follows fishing boats. Can dive clumsily from lower height. Often seen high in the sky, circling in loose flocks, or in direct flight on route to feeding or roosting sites. The usual call is a loud '**glaa**-o', much repeated. Courtship call is a series of 'laughing' notes with an initial drawn-out triumphant crowing, 'aau... ky**yaaah** kya-kya-kya-kya-kya-...'. Alarm call (caused by large raptor) a loud, annoyed 'kla-**aw** kla-**aw** kla-**aw**...'. **RSWP**

Yellow-legged Gull

Yellow-legged Gull *Larus michahellis* L 55, W 143. Breeds mainly colonially in the Mediterranean, but is also widespread on the Atlantic islands and along the Atlantic coast north to Bretagne. Scattered breeding sites on the Continent (e.g. Poland). Straggles widely after breeding and in winter, including to Britain and S Baltic. Very similar to closely related Herring Gull. Slightly different proportions: *wingtips protrude more* on standing bird, *legs are longer*, stout bill has *more marked gonys angle* and *more markedly down-curved tip*. Adult has *yellow legs* and *bright yellow bill with large red spot*, has a shade *darker grey upperwing, more black and smaller white spots on wingtip*. Head mainly white in autumn in E Mediterranean, but more streaked in the W and in the Atlantic. Iris yellow. Juvenile is less uniform than juvenile Herring Gull, has *whiter head* (with much dark around eye), *fore-neck and underparts* (with dark spots on flanks and sides of breast); *dark tertials white-tipped*, without pale indentations, outer greater coverts dark, especially at base, and unlike in Herring, pale 'window' of inner primaries is missing or very faint; *white rump/inner-tail contrasts to distinct, dark tail-band*, and *wings are rather dark below*, contrasting to whitish belly. Matures quicker than Herring Gull; mantle is moulted to new greyish, anchor-marked feathers as early as in Oct, and plain grey feathers can appear in first summer. Calls are more nasal and slightly deeper in pitch than Herring Gull, recall Lesser Black-back's. **WP**

Caspian Gull

Caspian Gull *Larus cachinnans* L 55, W 145. Breeds colonially or in smaller groups on coast of Black Sea (except in SW), Caspian Sea, at lakes and reservoirs in E Europe. Has expanded westwards, now breeds in C Poland. Like Yellow-legged Gull, straggles widely after breeding and in winter, is regular in Baltic and in W to Britain. Closely related to Herring Gull and Yellow-legged Gull, can hybridise with them where they meet. Has a more drawn-out, *sloping forehead* than Herring Gull, and the *bill is longer and more evenly thick*, with *less sharply down-curved tip* and *lacking marked gonys angle*. Further, *wingtips protrude* more on standing bird, and *legs and neck are longer*. In autumn, adults are picked out among Herring Gulls on whitish-looking head and usually *dark eye*, and identification is confirmed by shape of forehead and bill. At this time, bill is paler yellow (with a greenish tinge), and *legs are just 'pale'*, not yellow as when breeding. Juvenile has *whitish head and neck*, with an obvious *brown-streaked 'boa' on lower neck*, good contrast between *whitish rump and black tail-band, whitish underparts and underwing*, uniform *dark tertials* with broad whitish tips; greater coverts often dark basally, creating a dark band; on flight-feathers, a pale 'window' on inner primaries is hinted but not prominent. Calls similar to those of Yellow-legged. **WP**

adult

summer

winter (North Europe)

Herring Gull

winter

1st-winter

2nd-summer

adult

winter

adult summer

2nd-winter

Yellow-legged

1st-winter

adult

1st-winter

adult

2nd-winter

1st-winter

Caspian Gull

151

Audouin's Gull

Audouin's Gull *Larus audouinii* L 48, W 122. Breeds in small colonies on some islands in the Mediterranean Sea, and in a large colony at the Ebro delta. Distinguished from the most similar Yellow-legged Gull by slimmer build and by characteristic bill and by just *diminutive white dots at end of black wing-tips*. At a distance the dark red *bill looks black*. Note also that the grey on the back fades into white neck and tail without sharp division. *White trailing edge to secondaries narrow* and not prominent. Juvenile has whitish face and crown, all flight-feathers almost evenly dark, as are greater coverts. Tail all blackish with white tip. In second summer recalls Yellow-legged Gull of same age: pale grey back and wing-coverts, dark 'hand', dark trailing bar along 'arm', white tail with dark subterminal bar. However, *smaller* than Yellow-legged, and *tail bar narrower*. Call weak, hoarse and nasal.

Ring-billed Gull *Larus delawarensis* L 48, W 120. Rare visitor from North America, now annual in Britain (sometimes in small groups). Like a large Common Gull or small Herring Gull. Adults have yellowish *bill with distinct black band across*. Legs (greenish) yellowish, *iris pale*. First-winter similar to first-winter Common Gull but is larger, has heavier, *pink bill with black tip*, paler grey mantle, *more distinct spots* (some crescent-shaped) *on lower neck, breast and flanks*, and tail pattern differs: *dark subterminal tail band broken up* by paler narrow bars, especially distally (not all-black and clear-cut as in Common Gull); upper-and undertail-coverts rather prominently spotted. Legs pinkish. Second-winter told from Common Gull of same age by generally retained prominent dark spots on lower neck, by *remnants of narrow dark subterminal tail band*, by pale iris, and by bill and legs rather like in adult. **V**

Common Gull

Common Gull *Larus canus* L 43, W 109. Breeds commonly (in Britain locally) in isolated pairs or colonially, by coastal and inland waters, mainly in N Europe. Preferred nest site is high up, e.g. on large boulders sticking up out of the water (also on piles, even house roofs). Picks up worms on fields in flocks. Adult like Herring Gull, but smaller and narrower-winged, flies with quicker and more vigorous wingbeats, and has *dark eyes* and *weaker, greenish-yellow bill without red spot*. Distinguished from Kittiwake by prominent white patches on the otherwise black tips of the primaries, and on slower wingbeats. Juvenile has *sharply defined black band on the tail*, brown back which is moulted to blue-grey as early as the autumn, brown wing-coverts which are retained during the winter. Calls are higher and weaker than Herring Gull's, are loud, heard often, e.g. high cackling 'kakaka…', falsetto scream 'kleee-a', and also persistent alarm call '**klee**-u **klee**-u…'. **RSWP**

Mediterranean Gull

Mediterranean Gull *Larus melanocephalus* L 39, W 98. Nests colonially on steppe lakes, coastal marshes etc in SE Europe. Range expanding towards NW; occasionally breeds S England (can hybridise with Black-headed Gull), increasing in winter/on passage. Slightly larger and heavier than Black-headed Gull but smaller and shorter-winged than Common Gull. Adult distinguished from Black-headed by *heavier, more obtuse bill* (dark band near tip), by *pure white primaries*, as well as in summer plumage by hood which is black (not brown) and extends far down onto nape. When the hood is lost in late summer, the bird looks generally *very white*. Juvenile has brown back, but this is soon moulted to pale grey. First winter resembles Common Gull in corresponding plumage, but general size is *slightly smaller, back is paler grey* and wings more contrasting, having a *pure grey panel formed by greater coverts* between brown-streaked leading edge (lesser coverts) and *prominent blackish secondary bar*. Outer primaries blackish (with insignificant whitish subterminal spots showing when wings fully spread). *Ear-coverts blackish* forming a rather concentrated patch behind eye. *Bill and legs blackish*, too. Call nasal, somewhat recalling Arctic Skua, 'yeeah'. **SWP**

Audouin's Gull

adult

adult

juv.

juv.

Ring-billed Gull

2nd-winter

adult

1st-winter

1st-winter

adult

Common Gull

2nd-winter

adult

1st-winter

adult

1st-winter

2nd-winter

Mediterranean Gull

adult

1st-winter

adult

1st-winter

153

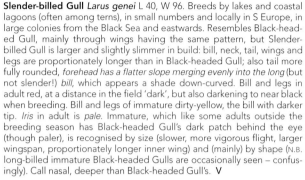

Slender-billed Gull

Slender-billed Gull *Larus genei* L 40, W 96. Breeds by lakes and coastal lagoons (often among terns), in small numbers and locally in S Europe, in large colonies from the Black Sea and eastwards. Resembles Black-headed Gull, mainly through wings having the same pattern, but Slender-billed Gull is larger and slightly slimmer in build: bill, neck, tail, wings and legs are proportionately longer than in Black-headed Gull; also tail more fully rounded, *forehead has a flatter slope merging evenly into the long* (but not slender!) *bill*, which appears a shade down-curved. Bill and legs in adult red, at a distance in the field 'dark', but also darkening to near black when breeding. Bill and legs of immature dirty-yellow, the bill with darker tip. *Iris* in adult is *pale*. Immature, which like some adults outside the breeding season has Black-headed Gull's dark patch behind the eye (though paler), is recognised by size (slower, more vigorous flight, larger wingspan, proportionately longer inner wing) and (mainly) by shape (N.B. long-billed immature Black-headed Gulls are occasionally seen – confusingly). Call nasal, deeper than Black-headed Gull's. **V**

Black-headed Gull

Black-headed Gull juv.

Black-headed Gull *Larus ridibundus* L 38, W 91. Nests commonly in colonies, which can sometimes become huge (thousands of pairs), on reedy lakes and marshes both inland and coastal, also on low islands. Colonies often contain breeding Tufted Ducks, Pochards etc, which derive protection from Crows. Often found in cities, where takes worms from lawns and the like, and in harbours, as well as on farmland. Black-headed Gulls benefit from cultivated landscapes; seen together with Common Gulls in large flocks following the plough. Catches winged insects in flocks high in the sky during the day, low over reeds in the evening. Winged ants a favourite food in calm summer days. The colonies are abandoned at end of July and most breeders from N Europe then move south to coasts and farmland, returning at end of March and waiting until their breeding sites are habitable. Britain's only abundant gull with *dark hood* (chocolate-brown). *Upperwings* in all plumages *have a triangular white panel* formed by the outer 4–5 primaries. *Underwings* are always *partly dark grey with a broad white fore-edge*. Adult has feet and bill dark brownish-red, in summer plumage brown hood leaving the nape white (a little white also at the eye), in winter plumage white head with dark spot at the ear. Juveniles are reddish-brown on back, nape and crown, soon moult this to grey and white, respectively, but retain brown wing-covert bar and dark tail band. Feet yellowish-brown, bill yellowish-brown with black tip. Slender-billed Gull is very similar in winter and immature plumages (see that species). Call a screaming, rolling 'krreeay', 'krre' etc. **RSWP**

Bonaparte's Gull juv.

Bonaparte's Gull *Larus philadelphia* L 33, W 82. Breeds in desolate coniferous forests in North America, in isolated pairs with the nests up in spruces. Rare vagrant to W Europe, most often in winter, often among Black-headed Gulls. In summer, adult is distinguished from Black-headed Gull by *sooty-black* (not brown) *hood* and *all-black bill*. Legs bright red. In all plumages *undersides of primaries are pale greyish-white*, not largely dark as in Black-headed. Smaller and more elegant than Black-headed Gull, especially in flight, which is buoyant and resembles Little Gull's. The immature resembles immature Black-headed, but the leading, longest primaries and especially the primary coverts have more black (see figs. at left); as in the full adult it lacks dark grey patch on underside of primaries. Bill grey-black, legs yellowish. Call a low, almost Coot-like cackle. **V**

adult

1st-winter

1st-winter

Slender-billed Gull

adult summer

juv.

adult winter

1st-winter

Black-headed Gull

adult summer

1st-winter

juv.

adult winter

1st-winter

adult winter

Bonaparte's Gull

adult summer

1st-winter

Sabine's Gull

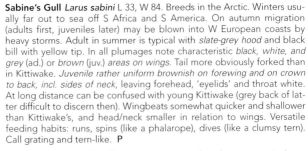

Sabine's Gull *Larus sabini* L 33, W 84. Breeds in the Arctic. Winters usually far out to sea off S Africa and S America. On autumn migration (adults first, juveniles later) may be blown into W European coasts by heavy storms. Adult in summer is typical with *slate-grey hood* and black bill with yellow tip. In all plumages note characteristic *black, white, and grey* (ad.) or *brown* (juv.) *areas on wings*. Tail more obviously forked than in Kittiwake. *Juvenile rather uniform brownish on forewing and on crown to back, incl. sides of neck*, leaving forehead, 'eyelids' and throat white. At long distance can be confused with young Kittiwake (grey back of latter difficult to discern then). Wingbeats somewhat quicker and shallower than Kittiwake's, and head/neck smaller in relation to wings. Versatile feeding habits: runs, spins (like a phalarope), dives (like a clumsy tern). Call grating and tern-like. **P**

Kittiwake

Kittiwake *Rissa tridactyla* L 40, W 95. Breeds Atlantic coast in large colonies on precipices of bird cliffs (the nests sited like swallows' on diminutive projections), sometimes also on conveniently situated building. Outside breeding season found mostly at sea. Swarms around fishing boats. Regular on passage, large numbers sometimes blown in to coasts in storms in autumn and winter. Flight then with *quicker and more mechanical wingbeats* than Common Gull's, in rough weather the action is more like Fulmar's. Tail square or shallowly forked. *Legs black* and rather short. Adults resemble Common Gulls but have *all-black wingtips* and upperside tricoloured: black wingtip, pale grey outer wing, darker grey inner wing and back. Seen against the light, outerwing appears thin and pointed. Immature with its black diagonal bar on upperwing recalls immature Little Gull and others but differs in broad, *distinct black nape band*, and also always *white crown* and pure white secondaries. 1st-summer generally lacks nape band and has smudgy dark markings on wing-coverts, can be confused with Sabine's Gull. Call, 'kitti**wee**ik', in nasal falsetto. **RSWP**

Little Gull

Little Gull *Larus minutus* L 26, W 63. Breeds in E Europe, occasionally in Britain, in reedy lakes, usually among Black-headed Gulls, and in tarns on taiga bogs. Sometimes in large numbers on passage. Maritime in winter. *Smallest* of Europe's gulls. In the evenings hunts winged insects over the reeds as Black-headed Gull, but flight considerably more rapid and elegant. Also snatches insects from surface of water. All year round, adults appear to have *rounded wings* with greyish-white upper side and *blackish underside with conspicuous white trailing edge*, in summer also a jet-black hood which extends far down onto nape. Juveniles have more pointed wings which are pale below and have a *black angled band above*. Back, mantle and crown are blackish-brown. Back is moulted to grey during first autumn, but dark upper mantle is retained somewhat longer, producing *Kittiwake-like look*. However, is markedly smaller, has darkish secondaries and often sooty-grey crown. Following spring/early summer black hood develops, angled wing pattern still shown. Calls loud and nasal, e.g. 'kep', often repeated in series. Display call 'ke-**kay** ke-**kay** ke-**kay**...' uttered with wings beating low down and neck upstretched. **WP**

Ross's Gull R*hodostethia rosea* L 30, W 77. Breeds in easternmost Siberia. Encountered in the pack-ice zone in winter and summer too. Very rare visitor to coasts of W Europe. As small as Little Gull (major confusion possibility), but has *longer and more pointed wings*. That tail is wedge-shaped (unique among Europe's gulls) is not so easily seen, but *the tail appears long*. Adult has *strong pink tone below* in summer and often a hint also in winter. Narrow pink necklace diagnostic. Wing markings essentially as in Little Gull, incl. *grey undersides with broad white rear edge*. Young has markings like young Little Gull. Flight light, tern-like. Silent outside breeding range. **V**

Sabine's Gull

adult winter

adult summer

juv.

juv.

Kittiwake

adult winter

adult summer

1st-winter

juv.

Little Gull

adult summer

adult
summer

2nd-winter

juv.

1st-winter

Ross's Gull

adult summer

1st-winter

adult winter

1st-winter

157

Laughing Gull *Larus atricilla* L 39, W 105. Very rare visitor from N America. Size of Common Gull but has *longer, more pointed wings*. Rather short-legged, gives long-bodied, slim impression on the ground. *Bill powerful and long, culmen curved down* giving 'drooping' look. Can be confused only with Franklin's Gull. Adult has black hood with white crescents above and below eye, sooty-grey mantle and upperwing (inner wing with white trailing edge) and *black wingtips without white*. First-winter is sooty-brown above, on breast and on the head (apart from pale forehead, chin and eye-ring), has very *dark flight feathers* (inner wing with white trailing edge), white uppertail-coverts and pale *grey* tail-feathers with broad black terminal band. Underwing-coverts and axillaries dusky and patterned grey. Blackish legs. Second-winter told from adults by more extensive black on wingtips and by traces of tail band remaining. **V**

Franklin's Gull *Larus pipixcan* L 34, W 87. Very rare visitor from North America. Slightly smaller than a Black-Headed Gull and has more rounded wings. A distinctive species, the only risk of confusion being Laughing Gull. In all plumages has *prominent white crescents above and below eye* ('swollen eyelids'), more so than in Laughing Gull. Bill rather heavy, but not so long and 'drooping' as in Laughing. Adult has black hood, *dark grey mantle and upperwing*, prominent white trailing edge, and *white area inside black and white wingtip*; centre of tail is light grey. First-winter has, like adult winter, a *dark half-hood* (darker and more clear-cut than in other winter gulls), rather *narrow dark tail band* leaving the outermost tail-feathers light (first-winter Laughing has wider tail band running across all tail-feathers), and pale underwing (Laughing has rather dark pattern on wing-coverts). Unlike other gulls has two complete moults a year, the second in early spring prior to northward migration: meaning that first-summer birds return with a more adult type of wing, though lacking the white area between dark wingtip and rest of upperwing; leading edge of primaries blackish, and dark hood is half or almost complete. **V**

Ivory Gull

Ivory Gull *Pagophila eburnea* L 44, W 107. Arctic species. Breeds in loose, small colonies, on shingly shores and on cliff precipices. Defends nest boldly. Patrols vast expanses of pack-ice and ledges of ice, seldom seen south of ice belt. Feeds on remains of seals killed by polar bears, on fish, carrion and offal, even on excrements from seal and polar bear. Snatches food from water surface but is reluctant to alight on the sea. Markedly fearless, readily resorts to camp sites. Said often to fly to places where guns are being discharged. A trifle larger than Common Gull. *Adult white, with yellowish bill with blue-grey base*. Immature is sparsely *spotted black above*, has darker bill and *'dirty'* face. The short *legs are black* in all plumages. The flight is light and elegant, wings being long with rather broad 'arm' and long, narrow 'hand'. Call recalls drake Wigeon's, 'pfeeoo', or with an 'r' sound, 'frreeoo'. **V**

Laughing Gull

adult summer

1st-winter

adult winter

1st-winter

juv.

Franklin's Gull

adult summer

1st-winter

adult winter

1st-winter

juv.

Ivory Gull

adult

adult

1st-winter

1st-winter

159

TERNS

Terns (order Charadriiformes, subfamily Sterninae)
Terns are adapted to fish or pick insects in shallow water and to migrate far between breeding and wintering ranges. They have slender build, long narrow and pointed wings for long-distance flights, and forked tail for maximum manoeuvrability during dives after food. Pointed bills, small feet and white, grey and black colours. Generally colonial breeders which can form quite large colonies (purpose: defense) in suitable areas. Clutches of 2–3 eggs. Several species move to the southern hemisphere during winter.

Black Tern

Black Tern *Chlidonias niger* L 24. Nests colonially in quagmire marshes and fens. Outside breeding season found mainly along coasts. Passage migrant in Britain, incl. inland. Feeds mainly by snatching insects from surface of water while moving about in graceful and playful flight. Often hangs fluttering few inches above aquatic vegetation. Also catches flying insects. Does not dive like the white terns, at least not during the breeding season. Unmistakable in summer plumage, which is *wholly dark* (dark grey, head and breast almost black) except for *pale grey underwings* and *white undertail-coverts*. Distinguished in winter and immature plumages from White-winged Black Tern mainly by *dark patch on sides of neck*, slightly *darker upperwings*, and as a rule grey, not whitish, uppertail-coverts (beware: occasional juveniles are paler here and in the field can appear almost white), as well as evenly grey tail. In addition the crown shawl is practically uniform dark grey. The commonest calls are a short, sharp, nasal, shrill 'kyeh' and a conversational 'klit'. **SP**

White-winged Black Tern

White-winged Black Tern *Chlidonias leucopterus* L 24. Breeds in SE Europe in swamplands. Seen on passage along sea coasts. Very like Black Tern in behaviour. Adult in summer plumage easily recognised even at long range by *jet-black body and underwing-coverts, gleaming white on the tail area and whitish on upperwing-coverts*. Compared with Black Tern usually has slightly shorter bill, broader wings and shorter tail – is slightly more compact, but the difference is subtle. Distinguished from Black Tern in winter and immature plumages by *lack of dark patch on sides of neck*, slightly *paler upperwings*, generally (but not always obviously) paler, whitish rump/uppertail-coverts, *whitish outer sides of tail* which contrast with otherwise light grey tail, and by paler crown shawl (front part is just dark marks on white background). In juveniles there is also *greater contrast between sooty-brown back* ('saddle') and *pale wings and tail*. (Winter plumage adults have rather pale grey back.) In winter plumage very like Whiskered Tern (which see). Call dry and harsh, 'kesch', like Grey Partridge in tone. Short conversational notes, 'kek'. **V**

Whiskered Tern

Whiskered Tern *Chlidonias hybrida* L 25. Breeds in swamps in S Europe. On migration on sea coasts. Similarity to Black Tern immediately revealed by restless, acrobatic flying around, usually low over the quagmire and flooded meadows, from which aquatic insects are snatched up (also dives from air). Plumage (adult, summer) appears at a distance whitish like *Sterna* terns, and in strong sunlight dark colour on underparts can be taken for effect of shadow. Still distinctive: *sooty-grey on whole of underparts* from belly to neck. *Head appears white except for the black tern-cap*. In winter plumage very like White-winged Black Tern, but can sometimes be distinguished by slightly *larger size, longer and heavier bill*, and no white wedge projecting up behind ear-coverts. Juvenile very like juvenile White-winged Black, but white *nape band narrow or broken* and white wedge projecting up into dark on ear-coverts is merely suggested. Mantle/back coarsely scaled sooty-brown/buffish-white (moulted early to pale grey first-winter feathers, but *tertials – typically blotched dark subterminally – are retained longer*). *Wings pale, lack markedly dark leading edge of 'arm'*. Rump whitish, tail pale grey with narrow dark subterminal band. Commonest call is a short, loud, crackling 'krsch'. **V**

Whiskered Tern juv.

White-winged Black Tern juv.

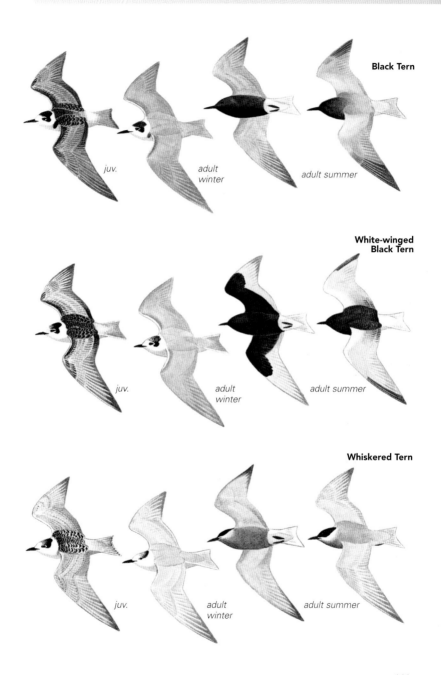

Black Tern

juv.

adult winter

adult summer

White-winged Black Tern

juv.

adult winter

adult summer

Whiskered Tern

juv.

adult winter

adult summer

Sandwich Tern

Sandwich Tern *Sterna sandvicensis* L 40. Nests in colonies locally on coastal islands, usually on sandy coasts. On migration wholly tied to sea. Fishes further out at sea, plunges from greater height than other terns. Often noticed by *call* and recognised by *general paleness, long, narrow wings*, short tail, powerful flight with deep rather hurried wingbeats, *long slender black bill* (with yellow tip) and, on ground, by suggestion of crest and black *short legs*. In winter plumage (often from July) forehead is white (not the whole crown as in Gull-billed, from which it is also distinguished by long slender bill and somewhat narrower wings). Juvenile usually has all-dark bill, white restricted to forehead (juvenile Gull-billed has most of crown white) and dark-patterned back and forewing coverts. The adults, which in late summer accompany their offspring during fishing, have contrastingly darker grey outer primaries and often different wing shape owing to moult. A species often revealed by its call, a loud, penetrating, grating 'ker**yick**', persistently repeated. Immatures' call is clearer, more ringing, 'sree-sri'. **SP**

Gull-billed Tern

Gull-billed Tern *Gelochelidon nilotica* L 38. Nests in colonies, uncommonly and locally in W and S Europe, on coastal marshes and sandy shores. Less tied to the sea than other terns, hunts mainly over land, e.g. coastal marshes and pasture meadows. Food consists mostly of insects – grasshoppers and crickets, beetles and dragonflies – but also of frogs, reptiles, mice and small crabs. Main species characteristic is the *gull-like, thick, black bill*. Leisurely, steady flight recalls Caspian Tern, but wings are very long, narrow and pointed, much as Sandwich Tern. No contrast between dark outer and pale inner primaries as in Sandwich. Dark trailing edge below to outer primaries narrow and distinct. Tail is short and shallowly forked. When standing, Gull-billed Tern shows characteristic *long black legs*. Rump and upper-tail pale grey. In winter plumage there is only very little black behind the eye (Sandwich Tern has black on the nape), *head* is therefore *almost all white*. Juvenile differs from juvenile Sandwich Tern but has whitish crown and a dark mark around eye, shorter and thicker bill together with different wing shape. Call a distinctive, nasal 'kay**wek**'. Alarm a rapid series of nasal, almost laughing notes, 'kevee-kevee-kevee-kevee', quite different from relatives (but resembling displaying Bar-tailed Godwit). The young beg with squeaky 'piuu' notes. **V**

Caspian Tern

Caspian Tern *Hydroprogne caspia* L 53. In Europe breeds mainly in the Baltic Sea, in large or small colonies on the outermost skerries of the archipelagoes, and feeds particularly in inner island groups and larger inland lakes (therefore flies daily tens of miles to and fro). Baltic Caspian Terns migrate partly across the Continent to the Mediterranean Sea on route to African winter quarters, do not avoid land like the Sandwich Tern. Dives for fish. Very big, *almost as large as Herring Gull*. Large, gleaming red bill ('carrot'), visible even at long range, is, together with *the call* and the *almost gull-like leisurely flight*, the best field character. In flight the head area appears strikingly large, *neck long and thick, projecting well forward*, and *tail short*. Legs black. Wing is pale grey above but has black wedge at tip below. Juveniles distinguished from adults by *more orange bill* with dark tip, dark brown cap (speckled white) which reaches farther down than on adult, varyingly dark *scaly pattern* on back and wings, *dark primaries above* and also pale legs. Call very deep, loud and heron-like in harshness, 'kraay-ap'. Begging call of young often given on late-summer migration is a squeaky 'slee-wee'. **V**

Sandwich Tern

adult

winter

summer

juv.

juv.

Gull-billed Tern

adult

winter

summer

juv.

juv.

Caspian Tern

adult

adult summer

juv.

juv.

TERNS

Common Tern

Common Tern *Sterna hirundo* L 35. Breeds in isolated pairs or smallish colonies on coastal marshes and islands, also on shores of inland lakes (incl. those with murky water). Frequently dives for fish. Best distinguished from rather similar Arctic Tern by following: *shorter tail streamers*, broader wings, *faster and more rigidly clipping wingbeats* (yet quite elegant, and when displaying exquisitely slow, elegant wingbeats, exactly like displaying Arctic Tern), larger head (*longer neck and bill*), longer legs. *Outer primaries above are slightly darker grey than inner ones*, and the transition looks like a 'nick' in the rear edge of wing (Arctic Tern uniformly pale grey). Wings not so transparent when seen from below (mostly only the inner primaries). The *dark orange-red bill nearly always has a black tip*. Juvenile has *grey secondaries* (with white tips; not all-white) and clearly marked *dark grey leading edge of wing* above. Back with heavy, wavy, sooty and buff barring. *Inner part of bill usually orange-toned* (not all-dark bill as in Arctic Tern). Voice often noticeably deeper than Arctic Tern's. Usual calls while fishing, courting and squabbling are short sharp 'kitt', rapid 'kyekyekyekye...' and also characteristic 'kirri-kirri-kirri...'. Alarm against humans a drawn-out '**kree**-ah', against crows a sharp 'ktchay'. **SP**

Arctic Tern

Arctic Tern *Sterna paradisaea* L 38. Breeds in colonies on islands, grassy dunes etc on clear-water coasts and also by small upland tarns. Shows less bias towards diving for fish than Common Tern, more inclined to snatch small animals from the surface of the water and to catch insects in flight. Best distinguished from similar Common Tern by following: *longer tail streamers*, narrower wings, even more elegant flight with *leisurely, springy wingbeats*, small head (*short neck and bill*), very short legs. Upperwings uniformly light grey, without 'nick' in edge. On wing seen from below against the light *all the flight-feathers transparent. Blood-red bill*, normally without black tip. Grey-toned throatside of neck, separated from black cap by broad white band. Juvenile is recognised by *almost white secondaries* (grey with white tips in Common Tern) and also *not such dark grey leading edge of wing* above; back has only hint of wavy grey barring; *bill all-dark*. Several calls rather like Common Tern's, but in courting and squabbling they have respectively squeakier and clearer ('pee-pee-pee-pee', clear ringing 'pree-e') or harder and rattling calls ('kt-kt-kt-krrr-kt-kt-') than latter. Alarm call against crows or mink a loud, sharp 'kleeu'. **SP**

Roseate Tern

Roseate Tern *Sterna dougallii* L 38. Breeds sparsely and locally in W Europe along coasts, often with Common and Arctic Terns. *Paler above* than both its relatives, appears strikingly *whitish*. Wings proportionately shorter, and *tail streamers* normally *very long and pure white. Bill long and black with red only at base. Legs longer* than in Common Tern. When standing shows *broad white inner edge of folded primaries*. At close range faint pink tinge visible on breast in spring. In winter plumage like Common Tern, but general impression is whiter, and long tail streamers are retained. Juvenile has restricted white on forehead, pattern of back as juvenile Sandwich, primaries without dark on tips, legs black. In flight usually immediately told on *quick, shallow, mechanical wingbeats*, resembles Little Tern more than Common/Arctic, looks very white. *Calls characteristic*, a rapid soft 'chu**wik**' and a broad, hoarse rasping 'zraaaach'. **S**

Little Tern

Little Tern *Sternula albifrons* L 23. Breeds sparsely in small colonies or in single pairs on flat, sandy or shingle coasts (on the Continent also at shallow, sandy lakes). *Smallest* of Europe's terns. *White forehead* in all plumages, *black-tipped yellow bill* (in spring and early summer) and *yellow legs* (all year). Juvenile is recognised chiefly by small size. Short-tailed. Flies quickly with fast wingbeats (rather like ringed plovers when it rushes about over the water). Gives butterfly-like impression when hovering above the water. Often dives excitedly over and over again, in rapid succession. Active and noisy, call a hoarse, shrill 'pret-pret'. **S**

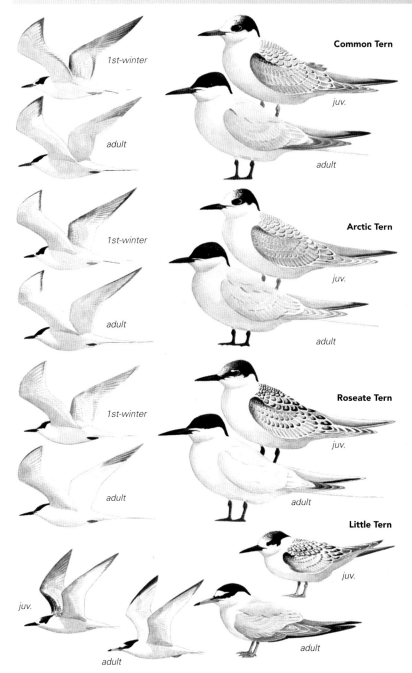

Common Tern

1st-winter

juv.

adult

adult

Arctic Tern

1st-winter

juv.

adult

adult

Roseate Tern

1st-winter

juv.

adult

adult

Little Tern

juv.

juv.

adult

adult

AUKS

Auks (order Charadriiformes, family Alcidae)
Black and white, sea-dwelling birds with short tails and narrow wings. Flight swift, close above the sea, wingbeats 'propeller-fast'. Silent outside breeding season. Come ashore only to breed. Nest in large colonies on sea cliffs (which look like gigantic beehives). Swim underwater using the wings, feet serve as rudder. Different winter and summer plumage. Immatures resemble the adults. Clutches, 1–2 eggs.

Guillemot

Guillemot *Uria aalge* L 40. Nests in large colonies (thousands) on ledges on vertical cliff faces along coast and on offshore islands. Lays single, pear-shaped egg directly on to narrow ledge. Parents recognise their egg by its appearance, their chick by its call. The young jump off more or less all at the same time at late dusk during a few evenings in July, still incapable of flight. The young bird is then guided on the open sea by the male. Heavy body, narrow wings, flies swiftly with propeller-fast wingbeats close over the sea, often several individuals in a line (Razorbill and Brünnich's do so too). Has fairly *long and slender bill*, differs from Razorbill in having longer, slimmer neck and shorter tail. Southern race (Ireland, southern Scotland southwards) dull blackish-*brown* on back, in intense direct sunlight looks dark grey-brown on back, a major distinction from jet-black Razorbill. Northern race (Scotland and N Europe) is darker, closer to Razorbill. Some individuals, known as 'Bridled' Guillemots, have a white ring around eye and a white line running backwards across top of cheeks. In flight neck is retracted; looks hunch-backed in comparison with Razorbill; seems to have larger hind body; feet well visible; *white on sides of rump much more restricted*; underwing-coverts not pure white, have some dark admixed. See also Brünnich's Guillemot. In winter plumage, distinguished from Razorbill and Brünnich's Guillemot by the fact that *white on sides of head extends higher up* and is divided by a dark streak running backwards from eye and also that *sides of body are streaked*. Call a rumbling 'a-orrr' slightly 'happier' in tone than Razorbill's. **RS**

Brünnich's Guillemot

Brünnich's Guillemot *Uria lomvia* L 39. Like Guillemot nests in large colonies on steep sea cliffs but has more northerly distribution. Best distinguished from Guillemot by *shorter and somewhat thicker bill* with thin *whitish streak at gape* (often hard to see) and also by *unstreaked body sides* (Guillemot always has some amount of dark brown streaking on flanks). Underwing-coverts and axillaries appear white. Also is *darker* brown-black above, almost black like Razorbill. In flight quite possible to pick out from Guillemot and Razorbill even at distance if background is light by following: *shorter, more compact, 'pumped up' body*; *hunch-backed*, even more so than Guillemot; bill pointing slightly downwards (Razorbill straight, Guillemot intermediate); *much white at sides of rump*, just like Razorbill; feet visible. In winter plumage white on chin and throat, but black covers a large part of cheek (not the case in Guillemot). Call rumbling, with 'malevolent' tone. **V**

Razorbill

Razorbill *Alca torda* L 38. Nests in cliff colonies (though not as large as Guillemot's). Lays single egg under boulders, does not require steep cliff face like Guillemot. Outside breeding season lives far out at sea. *Bill deep*, flattened sideways, is held raised as is the fairly long, pointed tail when Razorbill swims. Thicker and shorter neck than Guillemot. Distinguished from southern race of latter at long range by the fact that *head and back are jet-black*, not brownish. Underwing-coverts pure white. Razorbill differs from N Atlantic dark-backed Guillemots in having *more white on sides of rump*: white hind area with black central line, actually like Long-tailed Duck. Also, *holds head and tail higher*, does not look as hunchbacked as the guillemots, feet concealed by tail. Display flight startling: suddenly begins to fly in relatively slow motion. Call a grunting, jarring 'urrr', melancholy in tone. **RS**

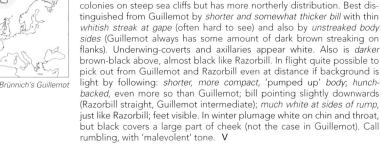

Guillemot

'bridled'

winter

Brünnich's Guillemot

winter

Razorbill

winter

167

Little Auk

Little Auk winter

Puffin winter

Little Auk *Alle alle* L 20. Exceedingly abundant in the Arctic, where it nests in enormous colonies in scree on the mountain slopes, not only at coasts but also inland on nunataks. The egg is laid in a burrow. One of the world's most numerous species. Flies over nest slope in swarms like mosquitoes. Very noisy, high-pitched trilling and nasal laughing calls echo in roaring chorus, 'keerrrr, kehehehehe'. Flies out to sea in flocks at high altitudes, fishes for plankton, returns in undulating bands, low over the water with chock-full throat sac. Outside breeding season lives far out at sea, but storms can blow them to coast (sometimes seeking shelter in harbours), even far inland. Much *smaller* than other auks, only size of a large Starling. Juvenile Puffin (underwing dark grey) is an obvious confusion risk where a single bird is seen flying past at a distance. *Whirring wingbeats, underwings dark*. Short bill. In flight appears extremely *short-necked*. On water usually swims with head retracted, looks quite neckless. Sometimes, though, stretches up and reveals surprisingly long neck. When resting, floats like a cork high on the water, glistens white, but when fishing, between dives, floats low, dark wings dragging in the water. **W**

Black Guillemot

Black Guillemot, juv.

Black Guillemot *Cepphus grylle* L 33. Nests in isolated pairs or small groups, locally forming small colonies, under boulders on rocky coasts. Less tied to the sea than other auks, lives all year round nearer shore. In summer, unmistakable with its *black plumage with large, oval white wing panels*. In flight, note short fusiform body, *rapidly whirring wingbeats*. Crimson-red feet glisten. *Always buzzes along low over the water*. In winter plumage mainly sparkling white, though white wing panel still darkframed. Juvenile is more grey-mottled, including scattering of grey even in wing panel (see fig.), on some individuals considerable. One-year-old birds may look all black (body feathers moulted, wing panels not). During the mating period Black Guillemots sit in groups and utter Rock Pipit-like 'seep-seep-seep-…' calls and drawn-out, thin but far-carrying 'electronic' peeps, at which time they open their bills wide (gape red). Also raise their wings and show the white wing panel and white underwing. **R**

Puffin

Puffin *Fratercula arctica* L 30. Nests in N. Atlantic, on rocky islands and on high rocky coasts, in colonies often containing thousands of pairs. Nest is in burrows in the earth or under boulder. In winter goes far out to sea, further out than any other auk except Little Auk. Note 'pot-bellied' body, big head and extraordinarily *deep, sideways-compressed bill*, grooved and gaudily coloured. At the breeding site birds are seen with their bill full of fish, which gives them a 'bearded' appearance. Frequently robbed of their fish by Arctic Skuas. The outer layers of the bill are shed in late summer, producing a somewhat less deep bill in winter dress. Immature has an even less deep bill. Greyish-white cheek darkens in winter plumage. When swimming, holds breast higher on the water than other auks, which lends characteristic silhouette. Usually seen in flocks. Flies low over the water in short lines. In flight *big pale head*, small size and short tail are striking. Large bill is pointed a little downwards. *Underwings fairly dark*, but not as blackish as Little Auk's. Call is heard only at breeding site and consists of unmusical 'aaah' notes, at the same time creaking and bellowing. **R**

winter, ready to dive

Little Auk

winter, resting

winter

Black Guillemot

1 year old

carrying fish

juv.

Puffin

adult winter

juv.

PIGEONS AND DOVES (order Columbiformes, family Columbidae)

Medium-sized, rather heavy birds with pointed wings and rather long tails. Feed on the ground. Flight swift and enduring. On taking wing a clattering wing noise is often heard, which serves as a warning signal. Can drink with the bill immersed in water (other birds take a billful and lean the head backwards). Clutches of 2 white eggs. Young are fed with a special liquid, 'pigeon's milk', produced in the crop.

Rock Dove

Rock Dove *Columba livia* L 33. Locally common in W and S Europe, in Britain now restricted to W coasts of Ireland and Scotland. Nests in mountain regions and often along coasts and on sea cliffs, where nest is made in caves or on sheltered cliff shelves. City Feral Pigeons (see below) are descended from this species and similar in appearance. Many Feral Pigeons markings almost exactly like the original: light grey upperparts but gleaming *white rump and two black bars on the wing. Underwings white* (Stock Dove's grey). Proportions and flight as Feral Pigeon, i.e. smaller, more compact in build than Wood Pigeon, also has faster wingbeats; flight very fast. Usually seen in small flocks. Cooing consists of series of muffled 'dru**oo**-u' notes, like Feral Pigeon's. **R**

Feral Pigeon *Columba livia* (domest.) L 33. Nests in towns and cities throughout almost all Europe, having originally escaped from dovecotes and become wild. Lives in towers, garrets, lofts and in niches on buildings, many broods per year. Fearless, feeds in streets and market places, may become very abundant. Plumage varies enormously, from the ancestral Rock Dove's grey plumage with *white tail base* and *two black wingbars* to *reddish-buff, white, strongly variegated white* or *dark individuals*. Flight and silhouette exactly like Rock Dove's. Resident, but at migration sites carrier pigeons (= Feral Pigeon) can sometimes be seen on 'migration'. Call resembles Rock Dove's. The young beg with drawn-out thin cheeps. **R**

Stock Dove

Stock Dove *Columba oenas* L 33. Fairly common and widespread in parks, farmland and edges of woodland, sometimes in ruins and on cliffs. Nests in holes in trees, buildings, on cliffs. Feeds on fields, often far from nest. Often visits shores. Feeds in small or medium-sized flocks, sometimes with Wood Pigeons. Is *smaller and more compact than Wood Pigeon*, flies with quicker wingbeats and, apparently, faster. Is richer grey than Wood Pigeon, *lacks white markings*, appears rather uniform in colour, but has *lighter ash-grey wing panel and lower back*. Traces of dark wingbars at bases of wings, not very noticeable. Whistling wing noise. Display flight with deep, well-spaced wing strokes and long glides on upraised wings (as Feral Pigeon). Call a monotonous fast croon, 'oo-e, oo-e oo-e,...', more like Feral Pigeon's than Wood Pigeon's. **RSW**

Wood Pigeon

Wood Pigeon *Columba palumbus* L 40. Most abundant and widespread of Europe's pigeons. Found in farmland, parks, gardens and all kinds of woodland. Breeds in flimsy twig nest. Has entered many larger towns and cities, where it mixes fearlessly with Feral Pigeons. Easily recognised by size, *white patches on sides of neck* (lacking in juveniles) together with *broad, white transverse bars on upperwings*. Outside breeding season usually seen in flocks, sometimes of considerable size. Makes loud wing clatter on rising. Flight with looser wingbeats than Stock Dove's and *tail* is proportionately slightly *longer* (though longer tail not always obvious; beware of some variation). Display flight is a short climb with wing clap at the peak, followed by short descending glide on half-closed wings. Muffled, slightly hoarse cooing consists of five syllables, with the emphasis on the first, '**doooh**-doo doo-doo, du'; final syllable is brief and abruptly cut short; series of notes is repeated three to five times without pause so that final, short syllable seems to belong to and to open next phrase. Also a growling, 'stomach-rumbling' 'hooh-hruh' **RW**

Rock Dove

Feral Pigeons feeding

Stock Dove

Wood Pigeons

juv.

Wood Pigeon

adult

171

Turtle Dove *Streptopelia turtur* L 26. Fairly common in S and C Europe in farmland with small woods. Summer visitor, winters in tropical Africa. Shy and watchful (much hunted). Clearly smaller and darker than Collared Dove, not so strikingly long-tailed. In flight easily recognised by *pale belly, fairly dark blue-grey underwings, small size* and *rapid, flicking wingbeats*. Has dark neck patch of black and white stripes (with touch of ash-blue). Juvenile lacks this. Tail dark with *broad white terminal band*, which stands out well when tail is spread. Red skin around eye obvious. Resembles Laughing Dove although this is an intrepid urban dove, which is clearly smaller, has Collared Dove proportions (short-winged, long-tailed) and has large blue wing panels (Turtle Dove's wing panels are smaller and paler grey-blue). Call is a deep, dry rumbling 'toorrrr, toorrrr, toorrrr'. **SP**

Turtle Dove

Collared Dove *Streptopelia decaocto* L 32. Since 1930 has spread to W Europe from the Balkans. First bred in Britain in 1955, now common and widespread. Closely associated with towns and villages, where it nests in parks and gardens. Lays several clutches, can be incubating in March and have young in nest in Nov. Often feeds in company with Feral Pigeons at silos, etc. Has nimble flight like Turtle Dove but is clearly stockier and is *markedly long-tailed*. Mainly *sand-coloured* with *narrow black collar* (absent in juvenile). Undertail white on outer half, black on inner. Uppertail sand-coloured with large white corner patches, concealed when tail is folded. A loud trisyllabic cooing, 'doo-**dooh**-do', is given from visible post, e.g. TV aerial. Also utters an almost mewing, 'annoyed' 'krreei'. **R**

Collared Dove

Laughing Dove *Streptopelia senegalensis* L 25. Recent immigrant to towns and villages on European side of the Bosporus. Not shy. *Long-tailed* and *short-winged* like Collared Dove but much smaller and *darker*, mainly *rufous with blue-grey markings*, more like Turtle Dove at first glance. Much white on outertail as latter, but has larger and darker blue wing panel and *characteristic chest markings* (black-spotted 'scarf'). Has rapid, rhythmic, subdued 5- or 6-syllable cooing, 'do-do-de**deedee**-do'.

Laughing Dove

SANDGROUSE (order Pteroclidiformes, family Pteroclidae)

Recall pigeons in shape and are closely related to them as well. Gregarious. Birds of steppe and deserts. Fly far to visit waterholes. Flight swift. Noisy in flight. Several species have capacity to transport drops of water in their belly feathers to their young waiting in the desert.

Black-bellied Sandgrouse *Pterocles orientalis* L 35. Inhabits arid plains on Iberian peninsula and in Turkey, and steppes at Caspian Sea. *Stocky build*, recalling small grouse on ground, pigeon when flying. Flight swift. Entire *belly black. Underwing-coverts white, flight-feathers black*. Note *lack of elongated tail-feathers*. Female has finely spotted breast and head. Flies far for water. Male carries water in soaked belly-feathers to young. Call characteristic: a far-carrying, snorting 'churrrl', recalling rolling snort from horse. A clear 'cheeeo', almost as from Little Owl, can also be heard.

Black-bellied Sandgrouse

Pin-tailed Sandgrouse *Pterocles alchata* L 30 (plus c. 7 for tail). Fairly scarce inhabitant of Iberian steppe country, local in S France. Short legs and mainly crouched posture, but neck can be stretched up considerably. Smaller and slimmer than Black-bellied Sandgrouse. *Central tail-feathers elongated* in both sexes. In flight recalls Golden Plover, then *appears neck-less*. *White belly* distinctive, and is also white on underwing-coverts. Plumage, rich in colours and patterns, provides efficient camouflage. Male *has black throat*, female not. May gather in large flocks outside breeding season. Call in flight is a nasal, far-carrying 'kyah, kyah,...' (or, when heard at closer range, slightly 'vibrating', 'kraow, kraow,...').

Pin-tailed Sandgrouse

adult

juv.

Turtle Dove

adult

juv.

Collared Dove

adult

Laughing Dove

Black-bellied Sandgrouse

♂

♀

Pin-tailed Sandgrouse

♂

♀

CUCKOOS

CUCKOOS (order Cuculiformes, family Cuculidae)
Medium-sized, long-tailed, with pointed wings. Two toes pointing forwards, two backwards. Some species (e.g. three European ones) are nest parasites.

Cuckoo

Cuckoo *Cuculus canorus* L 33. Fairly common in all types of terrain, in woodland and in open country right up to mountain slopes. Rarely in the immediate vicinity of densely populated areas, rather shy. Readily eats hairy larvae of bombycid moths. Nest parasite, lays its eggs in other birds' nests, one egg in each nest. Species often exploited include Pied Wagtail, Meadow Pipit, Whitethroat, Reed Warbler and Spotted Fly-catcher. Each female Cuckoo specialises in a particular host bird, whose egg colour it imitates. The newly hatched young Cuckoo pushes the eggs/young of the host birds out of the nest, getting benefit of all the feeding efforts of its host parents. Male Cuckoo is *ash-grey on head, breast and back, with Sparrowhawk-barring on the belly*. Female usually has same pattern apart from *rusty tint and suggestion of barring across upper breast* (grey phase), but a minority of adult females are instead *bright rusty-red above* (rufous phase). Juveniles are quite dark brownish above, some greyer, others more rusty, though not divided into two such different and well-defined categories as the adult females, do not become so bright rusty-red, are always darker brown-grey on head/neck. A sure sign of a juvenile is a *white spot on the nape*. Cuckoo's size, its low and unobtrusive flight progression combined with its *long tail* often give a Sparrowhawk impression (females of rufous phase: Kestrel). But quick wing-beats are fairly weak, and *the pointed wings perform without intervals of gliding* and *mostly below the horizontal*, and *small head* with delicate bill *is held pointing clearly upwards*. Often pursued by agitated small birds, e.g. Pied Wagtails. Male's call is the familiar Cuckoo-call, a far-carrying 'coo-koo', repeated in long series. Replaced by an irascible throat-clearing 'gugh-cheh-cheh' when male chases off another Cuckoo. Female has an urgent bubbling call, 'puhuhuhuhuhuhu', like calls of Whimbrel and Little Grebe. Begging call of the young is like that of a small bird, but very penetrating, 'sree, sree…'. **S**

Oriental Cuckoo

Oriental Cuckoo *Cuculus optatus* L 30. Shy bird of the taiga of E Russia and Siberia. Long-distance migrant (winters East Indies), could well over-shoot to W Europe (has appeared west as far as Latvia, reported even in Britain). Call disyllabic but obviously different from Cuckoo's: *both syllables at same pitch*, and *equal in stress*, besides being quite Hoopoe-like in tone, 'poo-poo'; repeated in series of 7–8, at *faster tempo* than Cuckoo's. The series is often begun (on alighting) with 5–7 'poo' calls in even, rapid succession. Otherwise depressingly difficult to identify: slightly smaller than Cuckoo, but usually has slightly bigger bill, thicker barring on belly (subtle difference), is a shade darker and colder grey above (crown, back), has rusty-yellow tinge on underparts more often than Cuckoo. The females can also be reddish-brown (black barring broader and more contrasting than in common Cuckoo). **V**

Great Spotted Cuckoo

Great Spotted Cuckoo *Clamator glandarius* L 40. Breeds in SW Europe in open country (scrubland, olive groves etc). Parasites mainly Magpie. Causes great alarm among Magpies when male tries to distract them and give female opportunity to lay egg. Young does not push young of host birds out of nest, but outfight them for food by superior begging behaviour. Adults return to Africa early in Jul–Aug. Resembles Cuckoo in shape and flight but is considerably *bigger*, has *crest*, has *broad white feather edges on otherwise quite dark upperparts*. Juvenile blackish-brown above, with rust-coloured primaries. Male's call, which is uttered frequently, is quite the opposite of Cuckoo's call, a rattling, loud cackle 'cherr-cherr-che-che-che-che-che', reminiscent of Turnstone or well-grown young of woodpeckers. **V**

juv. red morph

♀ *red morph*

♂

Cuckoo

young with foster parent (Meadow Pipit)

song posture

♂

♀

Oriental Cuckoo

adult

Great Spotted Cuckoo

juv.

OWLS (order Strigiformes, families Tytonidae, Barn Owl, and Strigidae, the other owls)
Owls have big heads and seemingly short necks, are birds of prey that hunt mostly at night. Both eyes directed forward, gives stereoscopic vision. Head can be turned up to 270°. Noiseless flight due to 'woolly' feathering. Sexes similar. Many species nest in holes. Some species make dangerous attacks in defence of their young.

Barn Owl

Barn Owl *Tyto alba* L 32. Thrives in open cultivated country, often nests in barns and church towers. Resident. Has greatly declined in N Europe. White/pale *heart-shaped face* typical. Flight and wing shape resemble Long-eared Owl, but *wingbeats quicker*, plumage is overall *very pale*, and *feet* often *dangle*. Birds in Britain and W and S Europe east to Italy pure white below and on face (race *alba*), whereas birds in Germany, Scandinavia and C and SE Europe are ochrous-buff below and on face (race *guttata*). Territorial call of male is a drawn-out (2 sec), hoarse screaming with a gargling quality. Has many other strong, hissing and snoring calls (some of which may sound 'spooky' by night), and a short yell is also heard. **R**

Scops Owl

Scops Owl *Otus scops* L 20. Breeds in S Europe in clumps of trees, gardens, also in towns. Nests in hollow in tree or building. Rather uniformly coloured, either brownish-grey or rufous-brown. *Small. Broad, 'thick-set' ear tufts.* Slimmer than Little Owl, and perches more upright. Wings rather long, feeds much on large insects. Truly *nocturnal*, noticed mostly by its call, a deep, short whistle, 'chook', monotonously repeated every 2–3 second. (Has been mixed up with call of midwife toad *Alytes obstetricans*.) Female's call is higher-pitched than male's, is used in duets. **V**

Pygmy Owl

Pygmy Owl *Glaucidium passerinum* L 17. Not uncommon in extensive coniferous forests. Nests in woodpecker hole. Europe's *smallest* owl, barely size of Starling. Hunter of small rodents, but also of tits and other small birds (in winter also at bird feeders in gardens). Often stores prey in nestboxes. More diurnal habits than most owls. Brown plumage *finely spotted white*. White supercilia angled, give *'stern' look*. Comparatively *small-headed*. Flight undulating as woodpecker. Territorial call is a soft, short whistle, 'hyuuk', rhythmically repeated every 2 seconds at dusk and dawn, mostly from very top of tall spruce. When excited inserts stammering 'tetete' between each call. Female may reply with higher-pitched, nasal version, 'hyeelk', or with thin, drawn-out 'pseeeee'. 'Autumn call' is a short series of shrill notes, rising in pitch, intensity and pace.

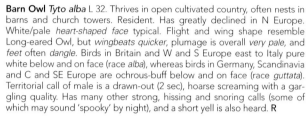

Little Owl

Little Owl *Athene noctua* L 23. Common in S and C Europe in often rocky open terrain, also in towns. Introduced in England in 19th century. Nests in holes in trees, rocks or buildings. Thickset body, *broad head with flat crown, long legs*. Posture usually not so upright as in the other owls. Bows and curtseys when agitated. Readily uses fence posts, telegraph wires, etc, as look-out. Flight undulating like woodpecker. Can hover. Prey includes rodents, birds, insects, earthworms. *Active by day* as well as night. Territorial call easily told from Scops Owl's, is more *drawn-out and rises at end*, 'gooo(e)k'. Repeated, loud, piercing mew '**kee**oo' is also often heard. Alarm shrill, explosive, tern-like 'kyitt, kyitt'. **R**

Tengmalm's Owl

Tengmalm's Owl *Aegolius funereus* L 25. Not uncommon in extensive coniferous forests. Nests in Black Woodpecker holes and Goldeneye nestboxes. Much larger than Pygmy Owl and different in shape (*large-headed!*), facial expression (*looks 'astonished'* with 'eyes wide open'!) and behaviour (true *nocturnal* owl, unobtrusive, avoids perch in top of tree). *Face pale but dark-framed.* Young are rich dark brown. Territorial call, heard mostly very early in spring, is a rapid series of short, deep whistles, usually 7–8 syllables, slightly rising in pitch at end, 'popopopopapa'. Carries far, up to 3 km. Remarkable individual variation in pitch, pace and number of syllables. Often calls persistently a good part of the night. A nasal, trumpeting 'ku-**wee**uk' can also be heard, as can (especially on autumn nights) a shrill smacking 'chee-**ak**' (squirrel-like!). **V**

Barn Owl

race guttata

race alba

young

Pygmy Owl

rustier

grey

Scops Owl

fledged young

Tengmalm's Owl

Little Owl

Eagle Owl

Eagle Owl *Bubo bubo* L 69. Local, mostly rare, in mountainous or rocky terrain, wooded or not. Shyer than other owls but appreciates elements of cultivated country (rubbish dumps). Resident. Nest usually on cliff shelf, at times beside rock on ground. *Largest* and most powerful of Europe's owls. Catches rats, voles, crows, gulls, ducks, even hares and birds of prey! Mainly nocturnal. Mobbed furiously by crows and gulls. Flight fast, with quite shallow wingbeats. Brown (colour of pine bark) above, *rusty-yellow below. Large ear tufts* (not visible in flight). Territory call mostly heard at dawn and dusk, a powerful '**hoo**-oh' (audible up to 5 km), repeated at, e.g. 8-sec. intervals (only first note heard at long range). Female replies with higher-pitched, hoarse version, but also has drawn-out hoarse bark '**ree**hew'. Alarm a shrill, nasal, fierce 'ke-ke ke**kay**u', with bill-clicking. Young beg with husky, scraping calls, 'chu**eesh**' ('planing wood'), into Sep.   **V**

Snowy Owl

Snowy Owl *Bubo scandiacus* L 61. Rare arctic owl, has bred in Shetland. Nests on high-lying, hummocky upland moor. Known as a diurnal owl but is mostly active after dusk. Uses mounds and rocks as look-outs. Relatively shy but male may draw blood when defending nest. Breeds in vole and lemming years, is absent in between. May migrate long distances south in winter, then kills larger prey, e.g. medium-sized birds. Stronger flier than most owls. Old males almost all-white, females (much larger) with fine dark spots, young males likewise, while young females are markedly more densely spotted (at a distance look dark grey against snow). Courtship call muffled far-carrying 'gawh', repeated at e.g. 4-sec. intervals. Male's alarm call is 'krek-krek-krek-…', strikingly like agitated female Mallard. Female's alarm call is a falsetto barking, 'pyeey, pyeey, pyeey', combined with a whistling 'seeuuee'. **WP**

Great Grey Owl

Great Grey Owl *Strix nebulosa* L 65. Uncommon breeder of the taiga. Large as Eagle Owl in external measurements but not even half as much in weight, and sticks to voles. Seems to be more sedentary than was previously thought. Nests in old raptor nest or on top of tree stump, i.e. *in view. Enormous head*, unmistakable face. Mainly *grey plumage*. Wings broader and more rounded than Eagle Owl's, flight almost in slow motion. Tail relatively long, coming rather to a point. Manoeuvres at leisurely pace but skilfully in dense forest. *Broad dark bar on tail tip* and large pale *rusty-yellow panel on undersides of primaries* diagnostic when flying directly away. Courtship call a series of 8–12 hoots, pumped forth at 0.5-sec. intervals, falling in pitch and intensity at the end, *extremely deep and muffled*, generally heard at most at 400 m. Female answers with surprisingly feeble 'chi**epp**-chi**epp**-chi**epp**' (a little like begging of large Tawny Owl young), also has drawn-out, extremely deep growling and grunting, low-voiced but penetrating alarm, 'grr**roooo**'. At the nest calm or aggressive but always fearless. Begging cries of young resemble female's call but hoarser.

Great Grey Owl   Ural Owl

Ural Owl

Ural Owl *Strix uralensis* L 57. Scarce breeder in taiga, in E Europe in deciduous forest. Hunts at clearings, forest marshes etc. Usually nests inside stormbroken dead tree ('chimney stack') but also in old Buzzard nests. Often aggressive at nest. (Do not approach young that have jumped out: they are defended with great determination and the owl aims for the eyes.) Mainly nocturnal. Resident. Much bigger than Tawny Owl but not so huge and imposing as Great Grey, resembles Buzzard when flying away. Tail relatively long, coming to a point, wings rounded. *Faded grey-brown. Face buff-grey*, unmarked. Courtship call deep, far-carrying: '**whoo**hoo … (4-sec. pause) … whoohoo o**whoo**hoo'. Gives muffled series of about 8 syllables 'poopoopoopoo…', gruffer than Short-eared Owl's and with slight rise just before end. Female has hoarse versions of these two calls and begs with scratchy, croaking, heron-like 'ku**veh**'. Alarm call like a dog's bark, 'waff'. Begging of young like that of Tawny Owl young.

Eagle Owl

juv. ♀

adult ♂

Snowy Owl

adult summer ♀

Great Grey Owl

attacking!

Ural Owl

179

Long-eared Owl

Long-eared Owl *Asio otus* L 34. Nests in old crow's nests in coniferous wood, often in clumps of trees in cultivated country. Not uncommon. Influxes from north in winter, when several may use same daytime roost. When alarmed adopts up-stretched posture, looking very much like stump of branch. Mostly nocturnal but can be seen hunting in daylight. *Wings long and rather narrow*, flight leisurely; resembles that of Short-eared, if not quite so persistently roaming about. In daylight teasing resemblance to flying Short-eared Owl; that the *wings are shorter and blunter* is not obvious; *finer and denser barring of wingtip and tail*, and *lack of white trailing edge to wing* often best features. Has drab mid-brown back and more uniform streaking below than Short-eared. *Iris orange*. Long ear tufts, not visible in flight, and can be lowered and difficult to discern even when perched. Has display flight with well-spaced wing claps (one per sec). Territorial call of male a deep hooting 'ooh' at slow-breathing rate, muffled (sounds very feeble, yet audible at 1 km). Female answers with a higher-pitched version or with relaxed, nasal 'paah'. Alarm 'kwek-kwek'. Begging call of young a drawn-out, mournful 'piii-eh', audible over 1 km. **RWP**

Short-eared Owl

Short-eared Owl *Asio flammea* L 35. A subarctic owl, numbers fluctuate with rodent cycles. Breeds locally in Britain. Nests on moorland and bogs. Winter influxes in south, then often found on coastal marshes. Resembles Long-eared Owl in flight but is more extreme, has *longer and narrower wings* than other owls, *beaten slowly* in rowing-like action with *hint of upward jerk*. Also active by day. Looks *pale in flight*. Breast heavily streaked, belly paler, almost unstreaked. Wing pattern similar to Long-eared's (obvious dark carpal 'comma' below, large dark carpal patch above) but *barring of wingtip is darker and coarser. Secondaries white-tipped. Tail has prominent and well-spaced bars* (much thinner and denser tail-barring in Long-eared). Iris yellow. Small ear tufts, revealed only when agitated. Male gives muffled courtship call, 'doo-doo-doo-doo-doo-...', in flight high up, difficult to pinpoint, sounds weak. Also short rattling wing claps (produced below belly, while losing some height). Female replies with husky 'cheeee-op'. Alarm call is a hoarse 'chef-chef'. Young beg with hissing, high-pitched version of female's 'cheeee-op'. **RSWP**

Tawny Owl

Tawny Owl *Strix aluco* L 42. Commonest and most widespread of Europe's owls. Found in mature woods, city parks and large gardens, both near and far from man. Resident. Nests in hollow deciduous trees or nest-boxes. Can actively defend downy young, aiming its claws at intruders eyes! Normal flight with series of rather quick wingbeats and long, straight glides. *Wings are broad and blunt, head is big*; gives all-together a compact impression. *Plumage rufous-brown, grey-brown* or something in-between. True nocturnal owl. Like a cat, hunts mainly by waiting on perch. Courtship call is the well-known 'hooooooh ... (4 sec pause) ... ho hoo-**hoo**'o'o'o' (final notes trembling) with tone of musical recorder. Female has hoarse, broken version, but also answers with cat-like 'kyu**weet**'. Tremulous trills ('bubbling call'), 'o'o'o'o'o'o'o'o'o'o'o', are heard from male, mainly at mating season. Territorial fights can be accompanied by weird cat-like screams. Alarm call is a rapid, explosive 'ku-**wit** ku-**wit**'. **R**

Hawk Owl

Hawk Owl young

Hawk Owl *Surnia ulula* L 37. A bird mainly of the subalpine region, especially in upper coniferous/lower birch zone. Nests in hollow tree. Often very aggressive at nest. Fluctuates markedly with voles, practically absent when prey is missing; can migrate far south then. *Longer tail* than other owls, and wings not so rounded. Very like large Sparrowhawk in *flight*, which *is swift and direct* with relatively quick and clipping wing action. Alights with steep ascent to elevated perch. Noted for being diurnal, but courtship call is heard mainly in pitch-dark night: a drawn-out and rapidly shivering trill (e.g. 95 syllables in 7 sec), broadly resembling Tawny Owl's bubbling call but more like female Cuckoo in tone. Audible at about 1 km. Female answers, and young begs, with drawn-out, hoarse 'ksheeeelip'. Alarm a shrill 'kvi-kvi-kvi-kvi-kvi', recalling Merlin but slower. **V**

Long-eared Owl

Short-eared Owl

rufous type

Tawny Owl

Hawk Owl

grey type

young

NIGHTJARS

NIGHTJARS (order Caprimulgiformes, family Caprimulgidae)
Noctural insectivores with big, flat heads, small bills and very big mouths. Stiff bristles at the corners of the bill help to catch moths in flight. The big, round eyes are kept virtually closed by day. Plumages brown, vermiculated and well camouflaged. During the day rest on ground or along branches and are difficult to detect. Two eggs are laid directly on to the ground.

Nightjar

Nightjar *Caprimulgus europaeus* L 28. Fairly local and scarce, found in light, dry, open woodland and in glades and clearings in denser coniferous wood, also on heaths. By day rests, hard to detect, lying flattened along thick branch, when finely patterned *brown and grey* plumage gives illusion of large flake of bark. In male, note *white spots on wings and tail*. Usually seen at dusk hunting insects. *Silhouette like Kestrel*. Flight silent and buoyant with stiff wingbeats. Often glides on wings held high, forming an obvious V, pitches and turns with great ease and elegance and sometimes stops still for a moment on fluttering wings. Seems almost weightless. Is rather curious, often makes a turn to have a look on intruders, sometimes approaching quite closely in hovering flight with body and tail hanging perpendicularly. Has an advertising flight with occasional wing clap (the wings are struck together above the body). Tiny bill but gigantic mouth – lives on moths, can even swallow cockchafers. Not gregarious except on migration, when flocks may occur. Migrates at night. Most often noticed by far-carrying song, heard without a break at dusk and night during early summer, a characteristic dry, hollow churring roll which runs in two gears, 'errrrrurrrrrrrerrrrrrr…'. When singing usually perches in the open, frequently aloft. Male courts female by pitching out and descending with rhythmic wing claps, uttering an intense 'feeoorr-feeoorr-feeoorr-…', which changes into a weak, deep and coarse 'ughrrrr…' (still producing wing claps), which in turn abruptly becomes silent – the whole gives a strong impression of 'the Nightjar's final sigh'! Call a frog-like, sonorous 'krruit'. **S**

Red-necked Nightjar

Red-necked Nightjar *Caprimulgus ruficollis* L 30. Breeds in Iberian peninsula. Found in evergreen woods, mainly of Stone Pine, and on dry, bushy waste. Night bird, closely resembling Nightjar, but bigger, with longer tail, slightly paler colours, *reddish nape band* and often slightly larger white throat patch (often divided by brown along middle of throat). White spots on wings and tail are distinct and are present in both sexes (cf. Nightjar), though fainter in the female. Much larger and darker in plumage than the rare Egyptian Nightjar. Male's song differs altogether from Nightjar's (but has something in common in its hollow ring) and is a protracted, often minutes-long series of repeated disyllabic 'kyotok-kyotok-kyotok-…'. Female gives a rasping, less far-carrying 'tche-tche-tche-…' ('steam engine'). Also wing claps. **V**

Egyptian Nightjar

Egyptian Nightjar *Caprimulgus aegyptius* L 25. Very rare visitor to S Europe from breeding areas in Africa and Asia. Found in desert areas, but usually near water. *Much paler* and appears more uniform in colour when perched than Nightjar and Red-necked Nightjar, which it resembles in behaviour and silhouette. Colours are perfectly matched to the dry clay of deserts, providing excellent camouflage to the resting bird. In flight, dark primaries contrast with otherwise faded upperparts. White markings of throat at times very indistinct. The song is something between the Nightjar's churring and the Red-necked Nightjar's rhythmic 'spelling out', a hollow-sounding 'kroo-kroo-kroo-kroo-….' in long series, in tempo like a slow old-fashioned motor in fishing boat. **V**

Nightjar

♂

♀

♂

♀

♂

Red-necked Nightjar

♂

hovering flight

♂

♂

Egyptian Nightjar

183

SWIFTS (order Apodiformes, family Apodidae)

Swifts resemble swallows and martins but have longer, narrower and more scythe-shaped wings as a result of an extreme adaption to life in the air. Feed exclusively on flying insects. Sexes alike. Nest under roof tiles, on cliffs and in cavities. Clutches of 2–3 white eggs.

Swift

White-rumped Swift

Little Swift

Pallid Swift

Alpine Swift

Swift *Apus apus* L 17. Common. Can be seen in the air almost everywhere but most often near towns and villages. Nests in small colonies, usually under roof tiles and in ventilation cavities, also in church towers; in wilderness in woodpecker holes. Lives on aerial plankton, gathered at heights of up to 4 km. Bad weather which makes feeding impossible in breeding area can cause mass migration to another part of the country, during which time the young fall into semi-torpor. *Uniform dull brownish-black* with pale chin. Cf. Pallid Swift. Is clearly larger than Swallow and House Martin, has longer, narrower, stiffer, scythe-shaped wings (extremely long hand and short arm) and very streamlined body. Flight outstanding. Standard flight fast with quick wingbeats (may give illusion of wings beaten alternately). Raids (often in screaming parties close above rooftops) exceedingly rapid, and accuracy allows Swift to enter nest almost like an arrowshot. But is often seen floating around quite leisurely high up in the air. Can 'sleep' during flight. Even mates in the air. Has difficulty in taking off from the ground, at least in tall grass. Call a shrill screaming 'srrreeee'. **SP**

White-rumped Swift *Apus caffer* L 14. Has established a few small colonies in S Spain since the 1960s. Nests in old nests of Red-rumped Swallow. Markedly smaller than Swift. Black with distinct pale chin, *narrow white rump and markedly forked tail*. Secondaries tipped white. Call a *jerky series of loud staccato notes*, 'cheet-cheet-cheet-…'.

Little Swift *Apus affinis* L 12. Inhabits towns and villages, also sheer cliffs. Has recently established small colony in S Spain. Distinguished from slightly larger White-rumped Swift by *square-cut tail* and larger *white, rectangular rump patch*. Call a *clear*, almost lark-like *twitter*. **V**

Pallid Swift *Apus pallidus* L 17.5. Fairly common in S Europe along rocky coasts, in mountain regions and in towns. Often seen together with Swifts and Alpine Swifts. Very like Swift. Overall plumage *slightly paler, more brown* and less sooty, which shows well from above in direct comparison. From below against the sky the colour difference is almost impossible to discern, but with experience one can often see that Pallid Swift's *inner primaries and secondaries are a shade paler* and more translucent; the *outer primaries are darker, creating a contrast*, which is slightly sharper than in Swift. Moreover, *dark eye-mask* ('sun-glasses') is more obvious, and *belly has faint scaly pattern* due to pale feather edges (note that young Swift also is somewhat scaly). Pallid Swift also has slightly broader wings, and as a result of this and its slightly larger size *beats its wings a trifle more slowly* in normal flight than Swift. It seems also to glide more than its darker relative. The *call is lower* than Swift's and a little *more harsh and strained*, 'vrreeeu', with a slight falling inflection, more so than in Swift. **V**

Alpine Swift *Apus melba* L 23. Breeds in S Europe. Found in mountain districts and cities (incl. at sea level). Nests in colonies. *Much bigger* than the other swifts and has *white underparts* together with *brown breast band*. Has slightly longer tail. Very fast and skilful in flight. Has clearly slower wingbeats than Swift – naturally, in view of its size – but maintains even faster speed. May even recall small, slender falcon. Just like Swift, is seen drifting around in large twittering flocks, especially in mornings and evenings. Call: rapid, *chittering series* 'titititi…', *rising and falling*, accelerating and slowing down again. Sometimes sounds rather hoarse as from small falcon. **V**

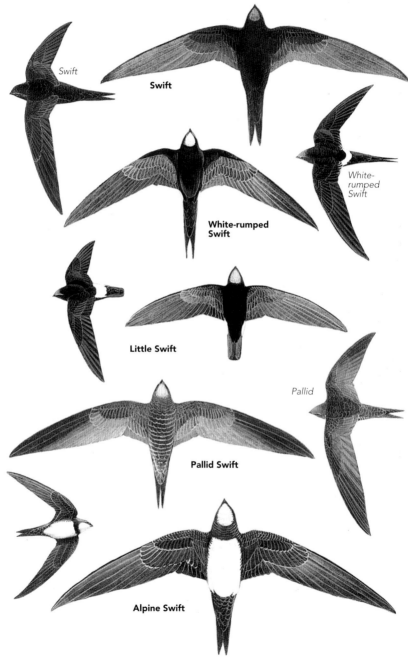

Swift

Swift

White-rumped Swift

White-rumped Swift

Little Swift

Pallid

Pallid Swift

Alpine Swift

KINGFISHERS AND ALLIES

Kingfishers and allies (order Coraciiformes)

are here represented by four colourful species belonging to four families.

KINGFISHERS (family Alcedinidae) are fish-catchers with large bills and small feet.

BEE-EATERS (family Meropidae) are skilful fliers, taking insects in flight. Long, thin bills.

ROLLERS (family Coraciidae) take insects on ground after watch from perch. Strong bills.

HOOPOE (family Upupidae) takes insects on ground when walking. Thin decurved bill.

Kingfisher

Kingfisher *Alcedo atthis* L 18. Breeds along calm rivers and streams with steep, clayey banks, where nest is excavated. Often seen at ponds rich in fish. Manages 2–3 broods. Large head, long bill, short wings, short tail and very short legs. *Shimmering blue and green above*, with back and tail appearing luminous. *Underparts reddish-orange.* Male's bill all blackish, female's has red base to lower mandible. Perches on branches overhanging the water, at bridges (often under them), etc, motionless for long periods, hard to detect (bright colours then not prominent). Dives head-first for fish, usually from perch, also after brief hovering. Rather shy. Flight swift and direct, close above the water. Little is then seen of the colours, but *back/tail usually gleam bright.* Even then difficult to catch sight of, but fortunately announces itself with high, piercing whistles, 'tzee!'. **R**

Bee-eater

Bee-eater *Merops apiaster* L 28. Fairly common in S Europe in open country. Nests in small colonies in burrows in sandpits or banks, or on flat ground, which it excavates itself. Colours of plumage bright and striking but not luminous. Fairly shy. Gregarious. Often perches on telephone wires. Catches insects in flight, often high up. In this recalls large House Martin: glides on outstretched pointed wings, flutters rapidly. In purposeful flight on the other hand rather thrush-like: thrush-sized, long-tailed, flight shallowly undulating. *Bill long, narrow*, slightly down-curved. *Central tail-feathers elongated.* Juvenile duller in colour, has mainly grey-green back, also lacks elongated tail-feathers. Call characteristic, quiet but far-carrying, hard to locate, 'klhut' (or 'krhut'), frequently repeated. **V**

Roller

Roller *Coracias garrulus* L 30. Breeds sparsely in S and E Europe. Declining species; range has contracted in recent decades. Prefers open, dry landscape with large hollow trees. Nests in holes in trees, banks, ruins, etc. The *pale blue of body and wings* has a luminous quality, in strong sunlight appears whitish azure-blue, in evening sun greenish-blue. Carpals and undersides of flight-feathers deep violet-blue. Juvenile is a little duller and browner. Perches in the open on dead branch or telegraph wire, flies down to take insect or lizard on ground. In flight somewhat resembles Jackdaw, but is faster with more vigorous wingbeats. Display flight includes a dive during which the male pitches from side to side (half rolls), like Lapwing. Calls '**chack**-ack' (like both Magpie and Jackdaw), 'rrak-rrak-rreh' (like angry Carrion Crow and Jay, respectively), and variants. **V**

Hoopoe

Hoopoe *Upupa epops* L 28. Fairly common in S and C Europe in open country with groves and cultivations. Nests in tree hole, walls, etc. Often seen near buildings but is rather shy. In Britain a regular 'overshooting' spring rarity (fewer in autumn), occasionally breeds. Sexes alike: *pale reddish-buff* with *striking black and white wing and tail barring*, erectile 'Red Indian chief crest' (raised when alighting, otherwise rarely), *long decurved bill*. Body size as large thrush. Owing to its broad wings appears larger in flight, which is flappy and desultory, resembling Jay's, often with irregular sweeps and flight path low down. In search of insects moves about mostly on the ground, where action is very 'vacillating'. The male's song is characteristic: a repeated, three-syllable, hollow 'poo poo poo', sounding weak at close range but still carries far. Usually performed when perched out of sight in canopy of tree. Also utters dry, Mistle Thrush-like 'terrr' and a harsh 'shaahr' recalling both Jay and Starling. **P**

Kingfisher

♂

hovering

Bee-eater

adult

Roller

juv.

crest up

Hoopoe

crest flattened

WOODPECKERS

WOODPECKERS (order Piciformes, family Picidae)
Woodpeckers are generally medium-sized birds. Feed partly on larvae of wood-boring insects, which are chipped out with their powerful bills. Use their stiff tail as prop when climbing trees (although Wryneck does not climb). Drumming has replaced song in most species (both sexes drum). All but Wryneck chip out their nests in tree trunks.

Green Woodpecker

Green Woodpecker *Picus viridis* L 34. Common in deciduous woodland with larger glades. Often seen on the ground searching for ants (hence camouflaging green back). In winter digs its way into anthills. *Green upperparts* with *brightly greenish-yellow rump.* Grey-headed Woodpecker has same colours, but Green differs by larger size, longer bill, *red on entire crown* (both sexes) and *more black on 'face'.* Male has red centre to else black moustache (Iberian race *sharpei* almost all red). Juvenile, in contrast to juvenile Grey-headed, is distinctly spotted on head, neck and underparts, and is red on crown. Shy and wary. Flight in long, marked undulations with well audible wing noise. Looks thin-necked then. Is voluble and noisy in spring and autumn. Song in spring is a rich, full-throated laughing series, 'kleu-kleu-kleu-kleu-…', accelerating and slightly falling in pitch at the end, used by both sexes (often in duets). Flight call is a shrill, short, vehement 'kyukyukyu**kyuk**', often repeated. Alarm call a more suppressed 'kyakyakya'. Begging call of young is rasping, like sandpaper on wood. Fledged young calls 'ky**a**' with Jackdaw-voice. Drums only rarely; bursts almost as fast as Great Spotted Woodpecker's but twice as long and surprisingly weak. **R**

Grey-headed Woodpecker

Grey-headed Woodpecker *Picus canus* L 29. Breeds across C Europe but is much less common than Green Woodpecker in the west. Tends to inhabit higher altitudes than latter. Also breeds in taiga (avoided by Green). In lowland accepts smaller woods, eg those fringing rivers. Ground-visiting but not so specialised on ants as Green; also feeds up in trees to a large extent. Shy and wary. In spring announces itself by frequent calls, in summer difficult to find, in winter visits bird feeders with tallow. Similar to Green Woodpecker but is smaller, has markedly *shorter bill, greyer 'face'* (only lores and moustachial stripe black). Male has *red just on forecrown,* female no red at all. Juveniles are not 'freckled' as juvenile Green, and just the males has a splash of red on the forehead. Song resembles Green's but the voice is clearer and weaker, almost fluting (easily imitated by whistling; often producing the bird), notes are more widely spaced, pace slows down at the end, and drop in pitch is more marked, 'kee-kee-kee kee kee kee kü ku'. Also calls 'kyuck' as the spotted woodpeckers and a quick hoarse "chechecheche' recalling Fieldfare. In contrast to Green Woodpecker drums frequently; rapid bursts, 1.5 sec long and considerably louder than Green's.

Black Woodpecker

Black Woodpecker *Dryocopus martius* L 44. Not uncommon in mature forests (often involving pines or beech). *Largest* of Europe's woodpeckers. Easily recognised by uniform *black plumage* and size. Male has entire *crown red,* female only rear part of crown. Flight flapping and slightly uneven, but not undulating like in other woodpeckers, most like Nutcracker's or Jay's. Flight call a far-carrying 'krrrree-krrrree-krrrree-krrrree-krrrree-…' (also serving as alarm). When perched utters characteristic drawn-out, loud and clear 'klee-ay'. In early spring a loud ringing 'kwee-kwee-kwee-kwee-…' has function of song (together with drumming), resembling song of Green Woodpecker but lacks acceleration or drop in pitch (keeps the same 'wild and crazy' ring to the end). In courtship encounters utters a sound like a clucking noise and nasal squeak run into one. Drums frequently; bursts long (1.8–3.2 sec, female's shorter), very powerful and 'articulate', like machine-gun salvo.

Green Woodpecker

♂ ♀ juv.

Grey-headed Woodpecker

♂ ♀ juv.

Black Woodpecker

♂ ♀ young

Great Spotted Woodpecker

Great Spotted Woodpecker *Dendrocopos major* L 25. Common breeder in both deciduous and coniferous woods, often seen also in gardens and parks (but is the most wary of the pied woodpeckers). Eats larvae of wood-boring insects which are chipped out from trunks and branches, but has a more varied diet than its relatives. In winter lives to a large extent on seeds from spruce and pine cones, which are pecked out at places known as 'anvils', i.e. cavities in stumps or tree trunks where the cones are wedged while being worked on. Undertakes eruptive migrations depending on the seed production. Also eats suet from birdtables. In spring sucks sap to some extent. Has *a large white panel on each shoulder, bright red vent/undertail*, unstreaked sides and also very little (male) or no (female) red on head – the male has *only a small red patch where the crown merges into the nape*. Therefore most resembles Syrian Woodpecker, which see. Juvenile, is more like Middle Spotted Woodpecker in having pale red vent/undertail and quite a lot of red on the crown (though not right back to the nape), but *lacks flank streaking* and has a *black band from bill to nape* (or almost to the nape). Call a high, short and sharp 'kik', may be pounded out in series at 1 sec. intervals; also very fast, in scolding series. Drums often, bursts typically short (0.4–0.75 sec), loud and *incredibly fast*, abruptly cut short. **RW**

Syrian Woodpecker

Syrian Woodpecker *Dendrocopos syriacus* L 23. Breeds in SE Europe in comparatively open country: parks, orchards, vineyards, poplar alleys etc. Great expansion northwest in Europe during 1900s. Very similar to Great Spotted, but distinguished by *lack of black line from moustache up to nape, behind cheek*. Also more red on hindcrown (male), whiter forehead, less white on outer tail feathers and Bullfinch-red undertail (instead of scarlet). Juvenile (red-crowned!) is distinguished from juvenile Great Spotted by the same characters, from Middle Spotted by moustachial stripe extending to bill. Flanks may be faintly streaked and barred. Also, some red tinge across breast on some. The 'kik' call is strikingly similar to the scolding call of a Redshank, 'gyp-gyp-gyp-…'. Drumming rapid, like Great Spotted's, but longer, about 1 sec.

Middle Spotted Woodpecker

Middle Spotted Woodpecker *Dendrocopos medius* L 20. Breeds in central and S Europe in mature deciduous woods, particularly of oak and hornbeam. Bill comparatively weak, used more as a probe than as a hatchet. Usually moves about in the tree crown. In spring drinks maple sap. Like Great Spotted has white shoulder panels, but also *red crown* (female a little duller red) in all plumages. Note that juveniles of Great Spotted (and Syrian) have red crown (although not *entire* crown), but Middle Spotted differs in having more white on sides of head and neck (moustachial stripe does not reach bill nor nape), *streaked flanks* and pale *pink undertail* without sharp border with *belly*, which is toned *yellowish-brown*. During breeding season calls instead of drums, has a nasal, strained, slowly delivered call, 'mjaik, mjaik, mjaik,…'. Also common is a whole *series of* 'kik-ük-ük-ük-ük-…' *calls at trotting pace*, first note higher-pitched than the others, 'kik,'. Drumming extremely rare, bursts weak as in Lesser Spotted.

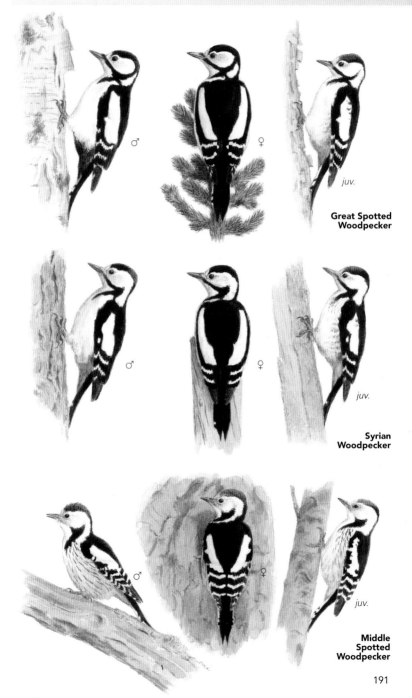

♂ ♀ juv.

Great Spotted Woodpecker

♂ ♀ juv.

Syrian Woodpecker

♂ ♀ juv.

Middle Spotted Woodpecker

White-backed
Woodpecker

White-backed Woodpecker *Dendrocopos leucotos* L 27. Rare, and over large parts of range fast retreating, because of its very particular habitat requirements: deciduous and mixed wood with ample supply of old decayed trees (especially aspen, alder, birch). Southern race *lilfordi* found in alpine fir forest. Often fearless. Often seeks food (larvae of tree insects) at ground level, e.g. on windfallen trees and in old stumps in centre of osier thickets. Bill long and powerful. Makes deep conical holes in rotten wood. Largest of the pied woodpeckers and has distinct white lower back and rump. On perched bird this is not always so obvious (Three-toed is more white-backed); *broad white transverse bar towards wing bend* is then the best mark. *Vent/undertail pale red*, fading out on belly and *streaked flanks*. Male has red crown, female all-black and juvenile not quite so much red as male. The 'kik' call is quieter and deeper than Great Spotted's, 'kük' or 'kok', more like Blackbird. Drumming powerful and c. 1.7 sec. long, resembling that of Three-toed's but *accelerating and fainter at end* ('shrinking', like diminishing bounces of a table-tennis ball).

Lesser Spotted
Woodpecker

Lesser Spotted Woodpecker *Dendrocopos minor* L 15. Breeds rather scarcely in deciduous and mixed woods, parks and subalpine birch forest. Particularly attracted to waterside areas with profuse tangles of foliage. Sometimes seen in orchards, in winter visits reedbeds where it pecks out pupae from the reeds. *Smallest* of Europe's woodpeckers. In its *deeply undulating flight* resembles an 'ordinary small bird', albeit a quite stocky one (Woodlark if anything). *Back strongly barred white*, can appear predominantly white. *Vent and undertail lacks red*. Commonest call very characteristic: a series of shrill, feeble notes, 'peet peet peet peet peet'. Also calls 'kik', though with weaker voice than its relatives. Drums frequently, but more slowly ('rattles') and more weakly than Great Spotted Woodpecker and for longer (c. 1.2 sec.). The sound volume of the bursts often varies; they are usually fainter in the middle and louder at beginning and end, may even be split by a fleeting pause. Characteristically, *the drumming is repeated at much shorter intervals* than in the other woodpeckers. **R**

Three-toed Woodpecker

Three-toed Woodpecker *Picoides tridactylus* L 22. Not uncommon in old spruce forest in the taiga zone. Also in subalpine birch forest. Scarce in mountain tracts of central and SE Europe, in coniferous forest. On average the most fearless woodpecker. Often feeds at low level. Habitually makes holes in a ring around trunk of large spruces to obtain sap. Pines, too, may be ringed. Head dark with white chin, *crown brass-yellow* in male, streaked black and white in female. Flanks very grey. Broad white 'blaze' right down back. The 'kik' call is usually soft and like a Redwing call, 'kyuk', but sometimes sharp and shrill like Great Spotted Woodpecker's 'kik'. Drumming powerful, longer (c. 1.3 sec.) than Great Spotted's and distinctly slower ('well articulated'), rather like short Black Woodpecker's.

Wryneck

Wryneck *Jynx torquilla* L 17. Fairly common on the Continent in open sunny woodlands (clearings), parks and orchards. Probably now extinct in England, but colonising NE Scotland. Arrives Apr–May, returns to Africa Aug–Sep. Not a 'real' woodpecker – does not climb trunks or chip away at wood, does not drum (in 'real' woodpecker manner). At first glance most resembles an ordinary small bird (large *Sylvia*). In tight situations, defends itself with snaking, twisting movements of neck (hence the name). *Plumage patterned like lichen in grey-brown with darker bands* along head and back. Tail long, bill relatively short. Specialised on ants. Spends most of time in trees, concealed in foliage, but also visits ground, hopping around. May be seen slipping away in low flight like a small, grey female Red-backed Shrike. Would be anonymous without its call, a series of loud, nasal, slightly croaky or moaning 'tee-e tee-e tee-e tee-e...', recalling small falcon. Begging call of young a rapid, high ticking 'tixixixixixix...'. **SP**

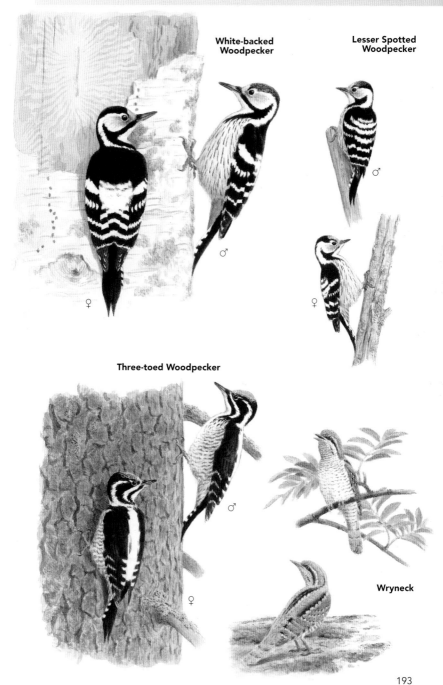

White-backed Woodpecker

Lesser Spotted Woodpecker

♂

♂

♀

♀

Three-toed Woodpecker

♂

♀

Wryneck

PASSERINES

PASSERINES (order Passeriformes)

Passerines vary in size from Raven to Goldcrest and form the largest of the orders, both in number of species and in number of individuals. All species have perching feet, with three toes pointing forwards and one backwards. In this group of birds with so many, often rather similar, species it is important to learn the characteristics of the families. Good guidance is given by bill shape, plumage colours and habits. The bill shape often indicates the feeding methods; insect-eaters usually have thin and quite weak bills, seed-eaters stout, conical bills. Calls are also important in identification.

LARKS (family Alaudidae) are small or medium-sized with mainly brown colours. They walk on the ground and are often seen in flocks, usually in open country. Heavier in build than the rather similar pipits. Nest on the ground. Number of species: 9. **p.196**

SWALLOWS AND MARTINS (family Hirundinidae) are small, long-winged, and have forked tails. Excellent fliers that catch insects in flight. Build characteristic nests, often in colonies. Number of species: 5. **p.202**

Woodlark

PIPITS AND WAGTAILS (family Motacillidae) resemble the larks in their terrestrial habits. The pipits are also brown, but slimmer and longer-tailed. The wagtails are brightly coloured and have long tails. Often seen in open country, sometimes in flocks. Number of species: 12. **p.204**

DIPPER (family Cinclidae) is a medium-sized, plump, essentially black and white bird which lives beside watercourses. Nests by water. **p.212**

WREN (family Troglodytidae) is a very small, plump, brown bird which lives in bushy areas. Builds domed nest. **p.212**

Swallow

ACCENTORS (family Prunellidae) are small, grey and brown birds, superficially sparrow-like but with thin bills. Shy. Found in woodland and also in bushy and mountainous country. Nest in bushes and on the ground. Number of species: 2. **p.212**

THRUSHES (family Turdidae) are medium-sized or small, mainly insectivorous species, spend much time on ground. Some species have very colourful plumages. The juveniles have spotted plumages. Includes wheatears, redstarts, Robin, nightingales, thrushes and many other species quite dissimilar to each other. Nest on the ground, in bushes, trees or cavities. Good singers. Number of species: 25. **p.214**

WARBLERS (family Sylviidae) are small, slim birds with thin bills and brown, green or grey plumages. Many species are very similar to each other and difficult to separate; song and calls are of great importance in identification. Found in dense vegetation, from reeds to woodland. Nest placed in dense ground vegetation or bushes. Number of species: 47. **p.228**

Stonechat

FLYCATCHERS (family Muscicapidae) are small birds with thin bills. Inhabit parkland and woodland. Catch insects in flight. Generally, hole nesters. Number of species: 5. **p.254**

TITS (family Paridae) are small, short-billed, agile birds with characteristic head markings. All are hole nesters; some species use nestboxes. Found mostly in woodland. Number of species: 9. **p.256**

Blue Tit

LONG-TAILED TIT (family Aegithalidae) is long-tailed and pale-coloured. Builds an intricate nest in a tree fork. **p.260**

BEARDED REEDLING (family Timaliidae) is a small, brown, long-tailed, tit-like bird that lives in reedbeds. **p.260**

PENDULINE TIT (family Remizidae) is a small, tit-like bird that builds a hanging nest **p.260**

NUTHATCHES (family Sittidae) have long and pointed bills and such strong feet that they can climb both up and down vertical surfaces. Nest in tree holes and rock crevices. Number of species: 4. **p.262**

Rock Nuthatch

WALLCREEPER (family Tichodromadidae) is a brightly coloured, round-winged bird that lives in rocky mountainous terrain. **p.264**

TREECREEPERS (family Certhiidae) are small and brown with long, decurved bills. Climb up tree trunks. Woodland birds. Nest in slits and under flakes of bark. Number of species: 2. **p.264**

SHRIKES (family Laniidae) are medium-sized, brightly coloured birds with 'raptors' bills' and long tails. Perch rather upright. They can be seen in bushy, semi-open country. Nest in bushes. Number of species: 6. **p.266**

Golden Oriole

GOLDEN ORIOLE (family Oriolidae) is a medium-sized, yellow and black (female greenish), thrush-like bird which is found in woods and parks. Open nest in tree fork. **p.268**

STARLINGS (family Sturnidae) are medium-sized, short-tailed and mainly black. Very sociable. Found in light woodland and in open country. Hole nesters. Number of species: 3. **p.268**

WAXWING (family Bombycillidae) is a medium-sized, greyish-brown and crested bird. Feeds on berries and insects. Lives in woodland. **p.268**

CROWS (family Corvidae) are large, often dark-coloured birds with rounded wings. Omnivorous. Often gregarious. Occur in most habitats. Nest in trees or on cliffs. Number of species: 12. **p.270**

Hooded Crow

SPARROWS (family Passeridae) are small birds, brown, grey and black in colour. Seed-eaters. Best known is the House Sparrow. Nest in cavities or build domed nest in trees and bushes. Sociable. Number of species: 5. **p.276**

FINCHES (family Fringillidae) have strong, conical bills adapted for seed-eating. Many species have colourful plumages. Sexes differ. Gregarious outside breeding season. Most live in woodland. Number of species: 22. **p.278**

Goldfinch

BUNTINGS (family Emberizidae) are medium-sized, brown, yellow and black birds with stout, conical bills. Most are found in open or semi-open country. In the majority of species the sexes differ. Partly gregarious outside the breeding season. Nest on the ground or in low bushes. Number of species: 15. **p.288**

Corn Bunting

Larks (family Alaudidae)

Small and medium-sized birds, rather uniform in colour, mainly brown. In Britain only two breeding species and one scarce winter visitor; most are found in S Europe. Inhabit open country. Rather like pipits, but stockier and with broader wings and shorter tails. Sexes usually alike. Good singers, usually during prolonged song flight at high altitude. Eat insects and seeds. Nest on the ground. Clutches of 3–5 eggs.

Lesser Short-toed Lark

Lesser Short-toed Lark *Calandrella rufescens* L 14. Breeds locally in SW and SE Europe in dry, open country, usually on more arid ground with less vegetation than Short-toed Lark. Very like latter, but adult is *distinctly streaked on the breast* and (more faintly) on the sides and is also usually slightly darker and *more greyish-brown*, has *shorter tertials which do not reach tip of primaries* on folded wing. Call a dry buzzing 'drrrr-drr' or simply 'prrrrt', like House Martin with dry ring of Sand Martin or like a scolding sparrow. Song can be problematically like Short-toed Lark's, though usually is *long, varied*, pleasing, *full of good mimicry*. Often it sounds like a Crested Lark, cleverly mimicked, changes over to trills like Calandra Lark's, strikes up Green Sandpiper, Common Sandpiper, Linnet and others, now and then relieved by the call. Often opened by repeated squeaky, piping notes when the bird is climbing. Song flight drifting around at random fairly high up, *wingbeats at times slow and deliberate* as in Greenfinch display – thus rather like Calandra Lark. **V**

Short-toed Lark

Short-toed Lark *Calandrella brachydactyla* L 15. Breeds in S Europe in open, dry country, often on dry mudflats. Like a small *pale* Skylark with proportionately slightly heavier bill. Clear *pale supercilium*. Usually rusty crown in fresh plumage. Dark-based median coverts stand out well against plain sandy lesser coverts, forming an *obvious dark bar*, like in Tawny Pipit. Underparts *unstreaked* except for a few, diffuse spots on the breast in the juvenile (adult plumage from Aug–Sep). Dark patch on side of neck in adult as a rule difficult to see in the field and may be absent. Very like Lesser Short-toed Lark, but adult distinguished from that species by lack of clear streaking on breast, generally paler and rustier-toned colours and often also, at close range, by dark patch on side of neck and *longer tertials* (covering tips of primaries). Call resembles Skylark's though more of a *dry chirrup*, 'drreet-it-it' or 'drri-e' etc. The song is composed of short, simple phrases, *little variation*, less often mimicking calls, mostly a dry chirruping similar to the call. *Second-long phrases, a good second between each*, phrases typically start faltering then accelerate. *Song flight undulating* in yo-yo fashion – Skylark-like *rapid fluttering wingbeats* succeeded by descending glide with tightly closed wings. Flight path during song wandering and erratic at high altitude. **V**

Calandra Lark

Calandra Lark *Melanocorypha calandra* L 19. Breeds in S Europe in dry, open country, mainly on natural or cultivated steppe, preferring grassy fields. Very *big* with proportionately short tail and *broad, long wings with white trailing edge* and typically *dark underwings* in flight. Bill thick. Large, *black neck patches*, less conspicuous in the female. Tail dark with white sides, like in Skylark. Flight usually low and undulating, wingbeats comparatively slow. Call powerful and *raucous, dry roll*, 'tshrreet' (like quarrelling Starling). The song, melodic and like Skylark's though more powerful and even richer in mimicry, is delivered during circling flight from high altitude, *wingbeats* for long periods *slow*, deliberate. In fair wind will at times stand still in the sky with deep, slow wingbeats, can then look peculiarly large (even like small raptor). **V**

Lesser Short-toed Lark

variation

Short-toed Lark

variation

Calandra Lark

*male can also sing from
ground with tail cocked*

197

Crested Lark *Galerida cristata* L 17. Fairly common in dry, open country with sparse vegetation, along roadsides and in open places in towns. Often seen close to grain silos, along embankments and, in coastal regions, in dock areas. Often very fearless. Runs fast. *Striking, pointed crest* (Skylark may sometimes raise crown feathers to a blunted crest), which distinguishes it from all other larks except Thekla Lark (which see). Bill fairly pale, *long and heavy*, pointed and 'fierce-looking', tip *straight or down-curved.* Rather pale brown (with slight ochre tone) with diffuse back markings. Looks *big-winged and short-tailed* in flight. *No white on trailing edge of wing*, and tail sides ochre. Juvenile has shorter crest and pale scaly pattern on back. Call clear, melancholy and a little languorous, e.g. 'dee dee düh'. Also utters hoarse mewing 'dwuee' and creaky, rolling 'drruee'. Song partly primitive, with pauses, combinations of clear call notes which are mostly delivered from perch, partly flowing, rich in mimicry, delivered in flight. **V**

Crested Lark

Thekla Lark *Galerida theklae* L 16. Breeds in SW Europe in dry, open country with sparse vegetation. Usually prefers rockier areas and higher altitudes than Crested Lark but also occurs in similar kinds of habitats. Very difficult to distinguish from Crested. A shade smaller, has *shorter and slightly darker bill with convex lower mandible* (Crested: straight), is usually greyer above and paler below, with *better-defined breast streaking* and grey-brown underwings (light brownish-pink in Crested). Uppertail-coverts usually rusty-toned, contrasting with greyer rump and tail feathers (only slight rusty tone and contrast in Crested), but this hard to see in the field. Often flies up into trees, which Crested seldom does. Call often of five syllables, 'tee-ti-tiuee'; can be shorter but always slightly softer (tone recalls Rustic Bunting song) and weaker than Crested's. Song resembles Crested's but has *softer, more melodious tone*, greater variation and complexity.

Thekla Lark

Dupont's Lark *Chersophilus duponti* L 18. Scarce and mainly resident breeder in C Spain and N Africa. Lives on dry steppes with grass or dry, low vegetation, e.g. low isolated tussocks. Famed for being utterly hard to get sight of, hides as a rail in the scanty vegetation, runs quickly away in cover when approached, is extremely unwilling to take wing. Stands more *upright* than other larks. Fairly *long, narrow and slightly decurved bill*. Crown has no suggestion of crest (but a hint of a paler central stripe). Pale supercilium and white outer tail-feathers. Rather short tail kept well folded, looks quite narrow. *Lacks white trailing edge on wings*. Pale-tipped upperparts feathers also in adult plumage (before worn) give variegated look. The song, which is performed in flight, high up with folded tail, is a brief, rather monotonous phrase, persistently repeated for minutes at a slow tempo, has a desolate ring and ends with a drawn-out, strained note, 'tru-tre-tru-**weeüüih**'. Sings mainly at late night, when stars are still out, but also in early morning. It is peculiarly hard to locate the bird on the song, since this appears to come from a spot on the ground even when the bird is 100 m above your head!

Dupont's Lark

Crested Lark

Thekla Lark

Dupont's Lark

Skylark

Skylark *Alauda arvensis* L 18. Very common breeding bird in open country, cultivated as well as natural (arable land, coastal meadows etc). Terrestrial. Rather pale brown, spotted dark, with *whitish trailing edge to wing* and medium-length white-edged tail. Has small crest. (Crested Lark has much longer and more pointed crest. Other rather similar species are Woodlark, Calandra Lark and Corn Bunting, which see.) Juveniles wear a scaly-patterned plumage until late summer, and lack pale trailing edge to wing. Usual call is a dry, full chirrup, 'prriee' and 'prreet' etc. The song is an *endless outpouring* based on high rolling notes repeated in long series, often containing notes of mimicry. It starts up as early as the first hours of dawn (mass starting up, strikingly simultaneous) and can then be heard all day. It generally begins on the ground or from a fence post but is usually performed for the most part high in the sky, for 10–15 minutes without stopping. The lark at this time hangs still on fluttering wings difficult to detect. Towards late summer the larks become silent. During the autumn loose flocks can be flushed on the stubble fields. In winter larger flocks are present, and in severe weather sometimes huge flocks and massive migratory movements may occur. **RSWP**

Woodlark

Woodlark *Lullula arborea* L 15. Scarce breeding bird on heaths with scattered trees, also felled woodland, burnt ground, occasionally nurseries. Resembles Skylark but has much *shorter tail*. Prominent pale supercilia which practically meet on the nape. Short crest which is usually not raised. Totally characteristic of the species is a *dark spot, enclosed by buffish-white, on the fore-edge of the wing* (upperside). Has a more distinctly undulating flight than Skylark, and seen from below it appears rather bat-like with its broad, rounded wings and obviously short tail. The tail lacks white outer feathers, instead has pale band at tip. The Woodlark often perches in trees and bushes (in exposed position on ends of branches, in the tops etc) but feeds on ground. The song consists of runs of mellow notes with a delightful ring, which begin tentatively but accelerate and increase in intensity at the same time as they fall in pitch 'lee lee lee lililililililulululoolooo, **ee**lu, **ee**lu **ee**lu-**ee**lu…'. It is heard mainly on early mornings and evenings, in June (2nd brood) sometimes at midnight – at which time it sings alone! Most often delivered in song flight, which is high up and often drifts away out of earshot, but also from perch (tree, telegraph wire, even on ground). Call a mellow, quiet, melodic 'deed-loo**ee**' or with a ring more like the song, 'tlo**ee**-tlo**ee**'. Small single-species flocks in autumn. **R**

Horned Lark

Horned Lark (Shore Lark) *Eremophila alpestris* L 17. In Europe an alpine species. The northern race (*flava*) has recently declined markedly in Fenno-Scandia, with a consequent reduction in numbers in W European wintering areas. Nests on dry mountain heath, often highest parts such as the top of low mountains. In winter found along sea coasts and on fields, on roughly the same ground as Snow Buntings. Easily recognised by *head markings of black and yellow;* feet black. Juveniles, however, have a completely different plumage: back, crown and cheeks dark brown with cream-coloured parts, supercilium and throat cream, broad brown-blotched breast band. Larks in Balkans and Turkey (*penicillata*) have more black on head (see plate) and more evenly grey upperparts. Distinguished in flight from Skylark by greater contrast between pale belly and blackish undertail. Flight like Skylark's. Usually, however, slips away low in Rock Pipit fashion. Call is a thin 'eeh tutu' (a squeaky note and two soft ones in rapid succession). Song short, jingling and irregular, resembles Lapland Bunting's and Snow Bunting's but more 'jolty', opens up almost like Corn Bunting, lacks drawn-out clinking notes of Lapland and mellow voice of Snow. Is usually delivered from a rock on the heath but also in circling flight, high up, with wings extended: tail then looks surprisingly long. **WP**

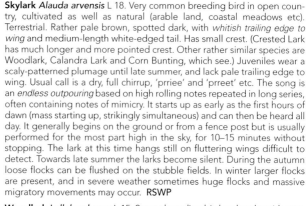

Skylark *adult*

juv.

Woodlark

Horned Lark

juv.

adult

Balkans ♂

201

Swallows and martins (family Hirundinidae)

Swallows have long pointed wings and often clearly forked tails. Their flight is swift and elegant with more varied and fluttering wing action than in the rather similar swifts. Legs and bills are short, mouths wide. Catch insects in flight. Often seen perched on telegraph wires, seldom on the ground. Long-distance migrants, the great majority leave Europe to winter in Africa. Often appear in large mixed flocks. Most species build characteristic sealed nests of mud, blades of grass and saliva. Clutches of 4–7 eggs.

Sand Martin

Sand Martin *Riparia riparia* L 13. Fairly common, colonial breeder in steep riverbanks, gravel-pits, sandpits etc, the nests excavated in the steep sides. Arrives in Britain mid Mar, a little earlier than the other swallows. Often seen near water. On migration, roosts in reeds in immense flocks. The smallest family member. *Upperparts brown* with no white. Characteristic of species is the *brown breast band*. See also Crag Martin. Call a low unmelodious rasp. **S**

Crag Martin

Crag Martin *Ptyonoprogne rupestris* L 15. Nests in colonies locally in S Europe in mountain regions and on rocky coasts. The Crag Martins of the Alps migrate south in winter, while Spanish breeders may stay behind. The nest is placed on cliffs. Larger and stouter than Sand Martin and *lacks breast band. Underwing coverts markedly dark*, forming distinct contrast with rest of underparts. *Wings broader* than Sand Martin's, and tail only slightly forked. Flight more restlessly agile and acrobatic than the other swallows'. When turning, the tail is spread, showing the *characteristic white spots*. Call a fairly weak 'tshree'. **V**

Swallow

Swallow *Hirundo rustica* L 19. Breeds commonly in cultivated, open country. Nests in single pairs or small, loose colonies. The open nest is made of mud and straw, usually inside buildings (on rafters, in recesses etc in barns and stables) or under bridges (similar siting). Roosts in large flocks in reeds on migration. *Outer tail-feathers greatly elongated and narrow.* Red forehead and chin good specific characters but difficult to see at some distance – look instead for *all-dark chin/throat. No white on rump.* Flight jerky and flicking (glides on extended wings are relatively short and fast, not as House Martin's). Call a short 'wit', 'wit-wit'. A fuller 'glit-glit' announces e.g. a Sparrowhawk. Sharp 'si-**vlit**' when mobbing. Song a rapidly delivered chatter with abruptly interposed wheezing sounds. **S**

Red-rumped Swallow

Red-rumped Swallow *Cecropis daurica* L 18. Fairly common in S Europe in open, preferably rocky country. The nest, with its tube-shaped entrance, is built entirely of mud and placed under projecting rocks, in the roof of a cave, under a bridge or on a building. Nests in single pairs or in small, loose colonies. Resembles Swallow, but note pale, *rusty-coloured rump and narrow rusty nape band, pale chin* and not quite such long tail projections (though long enough!). *Undertail coverts jet-black* (white in Swallow). (Note: hybrids between Swallow and House Martin can resemble this species.) Flight like House Martin's with glides on outstretched wings. Flight call a nasal 'tweit', rather like Tree Sparrow. Alarm a sharp 'keer'. The song recalls Swallow's but is slower, more grating and is often delivered in shorter phrases. **V**

House Martin

House Martin *Delichon urbicum* L 14. Nests commonly in colonies, mainly in villages and cities but also in mountain regions. The nest, in the shape of a bowl made by mud, is fixed under the eaves of buildings or on to cliff faces. Roosts in nests or treetops (not reeds). The House Martin is recognised by *blue-glossed black upperparts with white rump* together with white underparts. The tail is short and moderately forked. Flight more fluttering than Swallow's, with rather long, relaxed glides on extended wings (not flicking jerkily). Call a dry twitter, 'preet' (not rasping like Sand Martin's). Alarm call a repeated high 'seerr' (may mean Sparrowhawk, small falcon or – usually – internal quarrel). Song chirruping, on same pitch as the call. **S**

Sand Martin

Crag Martin

Swallow

adult

Red-rumped Swallow

House Martin

Pipits and wagtails (family Motacillidae)

Birds barely the size of sparrows, slender in build, with long tails with white outer edges and with slender, pointed bills. Lively and always on the go, very much terrestrial, able to walk and run well. Insectivorous. The wagtails have a habit of wagging their long tail up and down. The long tail serves as a balance in dashing runs. The pipits are brown and streaked and are best distinguished from the larks by more slender shape and longer tails. Clutches of 4–7.

Tree Pipit

Tree Pipit *Anthus trivialis* L 15. Fairly common (in places abundant) in open woods, in glades, at edges of clear-fellings and in more open country with bushes and trees. Resembles Meadow Pipit, but can be distinguished by slightly *more yellowish-brown colour*, not so grey-green, *belly whiter*, not yellowish-tinged, *more distinct dark lateral throat-stripe, flanks more finely streaked*, almost completely unstreaked rump, slightly stronger bill and comparatively short, curved hindclaw, but *best by voice*. Call a characteristic, hoarse 'speez'. Alarm a quiet but distinct, slowly repeated 'sit'. The song is sometimes delivered from a treetop but usually during a short song flight, in which the bird descends on stiff wings – like a little paper plane. The song is loud and is a repetition of short series of notes at varying tempo, 'cha-cha-cha-cha weeweeweewee trrrrrr uee uee uee uee, ooee **see**a-**see**a-**see**a'. **SP**

Olive-backed Pipit

Olive-backed Pipit *Anthus hodgsoni* L 14.5. Breeds in the taiga in NE Russia and Siberia. Rare autumn vagrant to NW Europe. Very like Tree Pipit, but is a shade smaller, more *olive-green above* with *fainter streaking on back; underparts* on the other hand are *more heavily streaked*. Supercilium broad and distinct, *vividly buff in front of eye*. Dark lateral crown-stripes. Also *a small pale and a small dark spot behind the ear-coverts* in most individuals. Rump unstreaked, and tertials have olive-grey edges with little contrast (Tree Pipit usually distinctly pale-edged tertials). Call a fine 'tseet', very like Tree Pipit's but a little less hoarse. From perch sings a Tree Pipit-like song but with less variation. Does not slow down with drawn-out '**see**a' like Tree Pipit, changes phrases more quickly, includes dry trills reminiscent of Red-throated Pipit. **V**

Tawny Pipit

Tawny Pipit *Anthus campestris* L 16.5. Locally common breeder in open, dry and sandy country with sparse vegetation, in S Europe sometimes on bare mountain heath. Annual in spring and autumn in Britain. *Sandy-coloured, rather faintly streaked*, above as well as below, and big, which distinguishes it from other European pipits. *Pale supercilium* and *dark lores* generally well marked, *median wing-coverts* dark, tipped light in fresh plumage. Juvenile in summer/early autumn is streaked above and spotted on breast like Richard's Pipit, but differs in weaker bill, *darker lores*, shorter legs and hindclaw and – above all – in *call*. Call very like House Sparrow's 'chilp'; some variation in articulation. Song a slowly repeated 'tseer**lee**', delivered in flight or from ground. Variations occur, e.g. a vibrant tremble descending in pitch: 'sr-r-ree-u'. **P**

Richard's Pipit

Richard's Pipit *Anthus richardi* L 18. A rare but annual visitor to W Europe, mainly at end of Sep and Oct, from breeding grounds in Asia. A few also winters in Iberia, Italy, Israel, etc. Often found on coastal meadows and dry long-grass fields. *Big and long-legged*, stands in *upright posture*, has *very long hindclaw*, streaked upperparts and a band of short streaks across breast. *Bill long* and stout, but can appear less striking on some. Most easily confused with Tawny Pipit, which also has streaked breast in summer and early autumn but Richard's differs in *call*, hindclaw, comparatively pale lores and often also in leg length and posture. Usually hovers immediately above ground before landing (which Tawny does only rarely). Call an explosive and loud, harsh 'schreep' on rising or in flight; at a distance sounds like a spluttery 'hissing' 'psch'. **V**

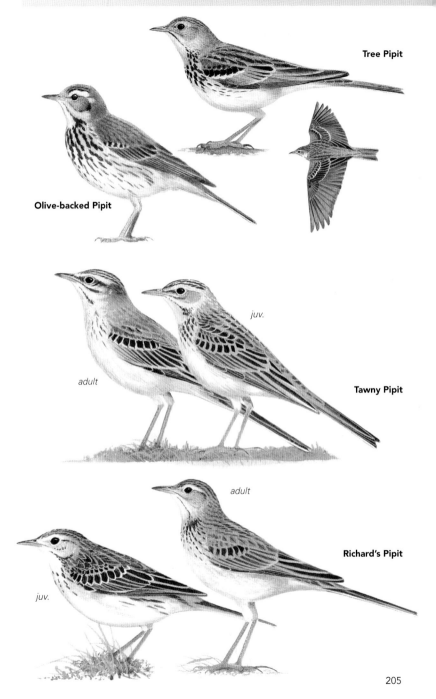

Tree Pipit

Olive-backed Pipit

juv.

adult

Tawny Pipit

adult

Richard's Pipit

juv.

205

Meadow Pipit

Red-throated Pipit

Water Pipit

Meadow Pipit *Anthus pratensis* L 14.5. Breeds commonly on moors, pastures, coastal meadows, dunes; abundant in N European mountains on heaths above tree line. In winter often found on fields, lowland marshes, coasts. Resembles Tree Pipit but a shade smaller, is rather *more grey-green*, has *less clearly defined lateral throat-stripe*, somewhat stronger flank-streaking, a trifle thinner bill and straighter, longer hindclaw. Rump very lightly streaked, clearly less so than back. Legs pink, pale. Rises with bounding flight and utters a thin 'ist-ist', 2 or 3 syllables, very unlike Tree and Red-throated Pipits but more like Rock Pipit's call (which see). Alarm a disyllabic trembling 'tirri'. Has brief song flight. Song simple: a few rapid series of sharp notes, e.g. 'zi zi zi-zi-zi-zi-zi-zu-zu zurrrrr seea seea seea seea seea seea seea', the final series of drawn-out notes delivered when the birds descends on stiff wings to the ground. **RSWP**

Red-throated Pipit *Anthus cervinus* L 14.5. Breeds sparsely in *Salix* scrub on subarctic bogs, mainly above tree line. Migrates south across E and C Europe, rare vagrant in west. In summer easily told from other pipits by *rusty-red throat, cheeks and upper breast*. Adults retain throat colour in autumn (although now a more subdued and yellowish shade). Juvenile resembles Meadow Pipit but has *heavily streaked rump*, is *rustier brown*, has two conspicuous *pale streaks on back* and prominent dark lateral throat-stripe. *Call* a very thin, drawn-out 'pseeh', distinctly different from Meadow and Tree Pipits. Alarm 'chup', like Ortolan Bunting. Song, often delivered in flight, most resembles Tree Pipit's but finer, has typical Redpoll-like dry trills and thin, drawn-out 'pseeeu-pseeeu-pseeeu- ...'. **V**

Water Pipit *Anthus spinoletta* L 16. Breeds in the mountains of C and S Europe, above tree line, on alpine meadows. In winter migrates to lower altitude, also to coastal areas further north. Occurs in S Britain on coastal meadows, by watercress beds, in reservoirs and by lakes. In summer plumage *crown and nape grey*, whereas *back, wings and particularly rump warm brown*, back very faintly streaked. Prominent whitish supercilium. *Unstreaked below* (or almost so), with more or less *pink tinge*. Pale wing-bar rather prominent. *Pure white on outer tail-feathers* (though exceptions recorded). Dark brown legs. Occasionally, male Rock Pipits in summer have underparts almost unstreaked pale pinkish-buff, but are distinguished by more uniform and olive-tinged upperparts and dusky outer tail-feathers. In winter, Water Pipits show a striking contrast between upperparts and underparts, compared with Rock Pipit: *grey-brown above*, with faint streaking, *whitish below*, with distinct and rather heavy streaking on breast and flanks. Pale supercilium, dark legs. Voice as Rock Pipit (which see). **WP**

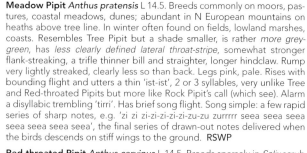

Rock Pipit

Rock Pipit *Anthus petrosus* L 16. Fairly common on rocky coasts of Britain and NW France; Scandinavian race *littoralis* occurs in Britain Oct–Apr. In summer plumage *upperparts rather uniform greyish-olive*, only slightly tinged brown (become greyer with wear), back faintly streaked; supercilium and wingbar *ill-defined*; breast and flanks usually well streaked, *streaks prominent but diffuse on yellowish-dusky ground; outer tail feathers marked dusky or greyish-white* (a little paler in *littoralis*). Variant male summer plumage described under Water Pipit. Distinguished from Meadow and Tree Pipits by *dark brown legs* (but juveniles in summer and autumn often reddish-brown), heavier bill and more diffuse streaking below on duskier ground. Song plain like Meadow's but differs on more rolling notes ('r' in voice) and rhythmic structure. Call like Meadow's though not so short and excitedly jumpy but more emphatic and usually only single or double note: 'weest' or 'weest-weest'. Alarm a Tree Pipit-like 'sit, sit'. **RWP**

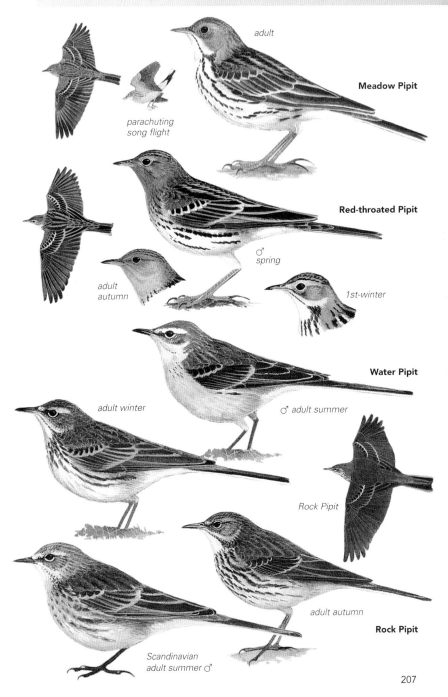

adult

Meadow Pipit

*parachuting
song flight*

Red-throated Pipit

♂ *spring*

*adult
autumn*

1st-winter

Water Pipit

adult winter

♂ *adult summer*

Rock Pipit

adult autumn

Rock Pipit

*Scandinavian
adult summer ♂*

Subspecies (or races) of Yellow Wagtail *Motacilla flava*

Subspecies	Male in summer plumage	Distribution
Motacilla f. flava	Blue-grey head, yellow throat, white supercilium	S Fenno-Scandia, W Europe except Britain and Iberian peninsula
M. f. thunbergi	Dark grey head, blackish ear-coverts, yellow throat, no supercilium	N Fenno-Scandia and Russia
M. f. flavissima	Greenish-yellow head, yellow throat, yellow supercilium	Britain and locally on adjacent Continental coast
M. f. iberiae	Grey head, white throat, narrow white supercilium	Iberian peninsula, S and SW France, Balearics
M. f. cinereocapilla	Grey head, white throat, generally no supercilium	Italy, central Mediterranean islands, Albania
M. f. feldegg	Black head, yellow throat, no supercilium	Balkans and Black Sea coast
M. f. beema	Grey (often pale) head, white supercilium and cheek-stripe	SE Russia
M. f. lutea (not illustrated)	Yellow head, yellow throat, pale yellow-green ear-coverts	Extreme SE Russia (lower Volga region)

The Yellow Wagtail complex of races is one of the avian systematist's hardest nuts to crack. Some prefer, e.g., to place *flavissima* and *lutea* in a separate species, *M. lutea*. Within one race's range individual males are often seen which resemble males of other races. These may be stray individuals of these races, but also simply colour variants of the race breeding locally. In the field one should be temperate in racial identification, especially as intermediate forms that are difficult to identify appear in border areas between the ranges of different races.

Yellow Wagtail

Yellow Wagtail *Motacilla flava* L 16.5. Fairly common breeder on damp meadows. Race *thunbergi* on delta-land, bogs, damp openings in birch forest on mountain slopes, vast clearings in coniferous forest. On migration often seeks company of grazing cattle, gets in close to hooves in search of insects. Flight undulating. Before migration often roosts in large flocks in reedbeds. Departs Aug–Oct, returns Mar–May. Long tail, *yellow underparts* and *greenish back* typical. Colour of head in summer plumage male varies depending on race (see table). Females in all regions are fairly alike. Some 1st-winter females are confusingly pale below, but usually have some yellow on vent. Juvenile plumage (moulted before southward migration) grey-brown above, dirty-white below, throat bordered by brownish-black upper breast band and lateral throat-stripe, pale supercilium with dark edges. Call a high 'pseet' or a fuller '**tslee**e'; race *feldegg* calls 'psrreet'. Song a mediocre 'srree-**srrit**' (or trisyllabic) with sharp scratchy tone, often delivered from bush top or barbed wire. **SP**

Citrine Wagtail

Citrine Wagtail *Motacilla citreola* L 18. Breeds on wet meadows in NE Europe, also in taiga, expanding westwards, has bred Finland, Poland. Vagrant in W Europe. Male unmistakable with *yellow head* and black hindneck, grey back and *two broad, pure white wingbars*. Female has grey-brown on crown and cheeks (latter entirely surrounded by yellow); at most a suggestion of black nape. The juvenile is grey above (faintly tinged brown, not green), has bold white wingbars, blackish uppertail-coverts, and sometimes slightly paler, brown-tinged forehead; below whitish, with buff tinge on breast (occasionally spotted), *undertail-coverts and vent white* (Yellow Wagtail: almost invariably pale yellow), flanks often grey like in Pied Wagtail. Tail longer than Yellow Wagtail's, about same as Pied. *Call on passage more piercing* than Yellow Wagtail's, which it resembles, more straight and *harsher* (hint of an 'r' sound), 'tsreep', can recall Richard's Pipit, but most resembles Yellow Wagtail song. **V**

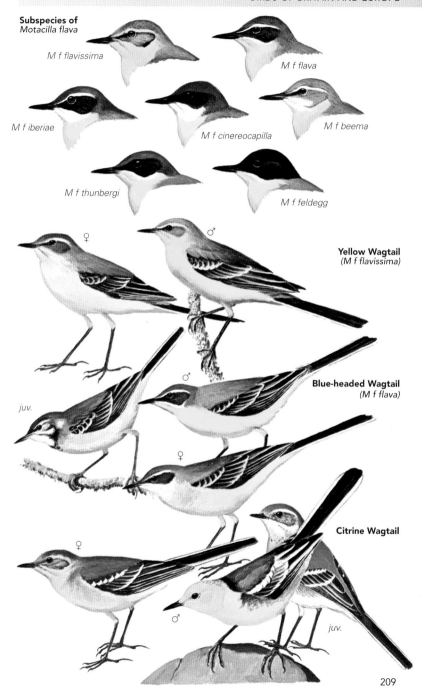

Subspecies of
Motacilla flava

M f flavissima

M f flava

M f iberiae

M f cinereocapilla

M f beema

M f thunbergi

M f feldegg

♀ ♂

Yellow Wagtail
(M f flavissima)

juv.

♂

Blue-headed Wagtail
(M f flava)

♀

Citrine Wagtail

♀

♂

juv.

209

Grey Wagtail

Grey Wagtail *Motacilla cinerea* L 18. Breeds beside upland streams and along swift-flowing brooks, sometimes beside lakes and slow rivers. Widespread and fairly common in S and W Europe, incl. Britain. The nest is placed on a rock ledge, under a bridge (often of stone) or in a similar site. Less particular outside breeding season; then also on coast, beside lakes, on sewage farms, watercress beds, cultivated land etc. Has *very long tail*, even longer than Pied Wagtail's. *Yellow on underparts*, especially *intense on vent, grey on upperparts* (white edging on tertials). Male's throat black in summer, female's faintly or profusely marked with black. When flying away *yellowish rump* and a *white wingbar* formed by white bases to flight feathers are visible. *Wingbar even more prominent from below, translucent* against the light. Immature is *yellow only on rump and vent*, the breast is buff. Distinguished from immature Yellow Wagtail by yellowish rump, extremely long tail, brownish-pink legs (blackish in other wagtails). White wingbar is also present in the immature. The long tail makes its mark on all the Grey Wagtail's movements; *the flight is even more markedly undulating* than Pied Wagtail's, the action on the ground *even more rocking and see-sawing*. Runs and skips deftly among the rocks of the rushing waters, often hovers above water in search of insects, is very much inclined (more than Pied Wagtail) to perch in tops of overhanging trees. Usual call resembles Pied Wagtail's but is markedly more metallic and higher-pitched, 'tsiziss'; this, together with wingbar, extreme tail length and extremely undulating flight, is what gives it away in overhead flight. Alarm 'sü-eat', mixed with excited version of call. Song consists of a short series of repeated sharp notes. **R**

Pied and White Wagtail

Pied Wagtail *Motacilla alba yarrellii* L 18.5. British Isles race of White Wagtail, also breeds sporadically on adjacent Continental coasts: locally in Norway, Germany, Holland, Belgium and NW France. Breeds commonly in open country, around farmyards, in towns, usually near water. Nests in recesses or holes, often under roof tiles and in stone walls, under stones, even on moored boats. *Black, dark grey and white plumage* together with *constantly wagging tail* distinguish it from all other birds. Male can often be told from female by back being pure black; female usually has dark grey back. Juvenile is grey on face and has a grey patch across breast. At end of summer moults into 1st-winter plumage with dull yellowish face and *prominent black crescent on upper breast*. Adult in winter similar but is white on face and throat. Juvenile has rather unmarked greyish head and a dark patch across breast. Birds in 1st-winter plumage and females in 1st-summer plumage resemble White Wagtail (see below), but *rump is almost black*, not grey, and *flanks are dark grey* (with green tone), not pale ash-grey. Outside breeding season usually seen in small parties, but sometimes gathers in large flocks to roost communally in reeds, orchards etc. (at times even inside factories and glasshouses). Feeds on insects taken on the ground as well as captured in the air in flycatcher fashion. Runs very quickly. Wags tail up and down and nods head as it moves. Flight deeply undulating. Call a disyllabic, kind of 'rebounding', 'tsee-**litt**' and variants. Song unobtrusive, a few notes in combination, varied and well paused. When agitated utters an alternative song variant, a prolonged twittering, sounding very excited and lively; also heard when chasing off Cuckoos and smaller birds of prey. **RSWP**

White Wagtail *Motacilla alba alba* L 18. Nominate race of Pied, breeds throughout Europe except within Pied's range. Male easily told from Pied Wagtail by *medium grey back* contrasting sharply with pure black nape. Female similar but grey back colour usually merges imperceptibly into black of crown. Juvenile impossible to distinguish in field from juvenile Pied (but 1st-winter birds often separable – see above under Pied). Calls as Pied. **P**

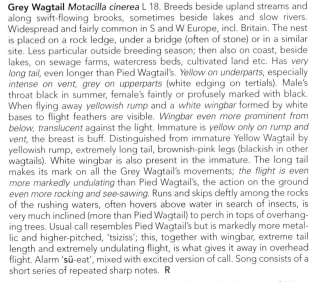

Grey Wagtail

♂ *summer*

♀ *summer*

juv.

♀ *winter*

♂ *winter*

♂ *summer*

Pied Wagtail

♀ *winter*

♂ *winter*

White Wagtail

♂ *summer*

juv.

Dipper (family Cinclidae)

The Dipper is, because of its powerful feet, solid skeleton and special oil gland, well adapted to a life in water. Sexes alike. 4–6 eggs, laid in large, domed nests in association with running water.

Dipper

Dipper *Cinclus cinclus* L 18. Nests along rapid-flowing water-courses, mostly in more hilly areas. In winter thinly distributed along rivers, streams and forest brooks, but a hardy species, will remain in areas where rivers are half-frozen. Perches on boulders etc., bobbing and curtseying, short tail slightly cocked. *White breast* conspicuous. Hops in the water, swims (sits low), dives for worms etc., swims underwater with powerful wing-strokes. Can even run along on the bottom. Also in mid-winter sings its quiet, squeaky, scratchy song. Flight is low and quite straight and fast. Then utters short, harsh 'zrets' which penetrates through the roar of the torrent. Continental birds are blacker below, lacking chestnut band, occasionally visit E coast of Britain in winter. **RW**

Wren (family Troglodytidae)

A small, very active, brown bird. The finely barred tail is usually held cocked. Sexes alike. The 5–7 eggs are laid in a large, domed nest.

Wren

Wren *Troglodytes troglodytes* L 10. Occurs commonly in a wide variety of habitats, in woodland, gardens, reeds, upland moors, cliffs etc. – anywhere with good low cover, incl. on barren islands far offshore. *Diminutive, cocked tail* and *rusty-brown plumage* distinguish it. (Several island races off N and W Scotland are larger, vary in strength of plumage colour.) Lives near the ground, often keeps well hidden. Call a dry rolling 'zerr'. Alarm a loud metallic 'zek, zek…'. Song melodious and unexpectedly loud, consists of rapid series of high, clear notes and trills, e.g. 'tee lu ti-ti-ti-ti-ti turrr-yu-tee-lee zel-zel-zel-zel-zel yu terrrrrrrrrr-zil'. In Britain (and elsewhere) sings loudly even in mid-winter. **RW**

Accentors (family Prunellidae)

Small, grey and brown, thin-billed, rather skulking birds. Feed on the ground. Mostly seen singly. Sexes alike. 4–6 blue-green eggs.

Alpine Accentor

Alpine Accentor *Prunella collaris* L 18. Breeds in central and S Europe on high mountains above the tree line. Lark-sized. At a distance looks dull brownish-grey, almost like a Water Pipit (which often shares its habitat, alpine heaths). The character visible at greatest range is the *dark bar across the wing* (created by the greater coverts), and next the *heavy chestnut markings on the flanks*. The finely barred throat markings are apparent only at close range. Yellow base of bill. Sexes alike. Gives quite loud, rolling, lark-like 'drrrüp-drrrüp-…' and thrush-like 'chep-chep-chep-. . .' calls. Song something between Dunnock and Shore Lark, tuneful and irresolute chirruping. Also delivered in song flight. **V**

Dunnock

Dunnock (Hedge Sparrow) *Prunella modularis* L 15. Common breeder in denser gardens and parks, also among junipers in cultivated areas, in scrubland, preferably half-grown spruce forest and subalpine birch forest. Note *slate-grey head and breast* together with streaked warm brown upperparts. Juvenile (seen in summer only) lacks slate-grey, is rather nondescript, streaked brown-black and yellowish-brown above and below. Feeds much on ground. Skulking, but song is delivered from exposed perch such as top of bush or very top of spruce. Song characteristic, a high-pitched, clear jingle, somewhat irresolute and markedly cyclic, 'tutel-lit**ee**tell**ee**t**ee**tut**e**llit**u**tell**i**t**ee**'. Common call (excitement, alarm) is a loud piping with a cracked tone, 'teeh'. Flight call (migration) a thin and frail ringing, 'sissississ'. **RW**

Dipper

northern race

British race

juv.

Wren

Alpine Accentor

adult

Dunnock

Thrushes, wheatears and allies (family Turdidae)

The true thrushes are dealt with on p. 224. The others follow here, with one exception (Rufous Bush Robin, p. 246). They are a varied bunch of smaller birds divided into several genera: wheatears (*Oenanthe*), chats (*Saxicola*), rock thrushes (*Monticola*), redstarts (*Phoenicurus*). Bluethroat and nightingales (*Luscinia*), robins (*Erithacus*, *Irania*) and Red-flanked Bluetail (*Tarsiger*). In several species, the males are prominent songsters. Juveniles are spotted. Build nests in low trees, bushes or in holes. Clutches of 4–7 eggs.

Redstart

Redstart *Phoenicurus phoenicurus* L 14. Breeds fairly commonly in parks and open forest, in suburbs as well as in remote taiga. Nests in tree holes and nestboxes. Male in summer attractive in black, white, ash-grey and orange-red. In Sep, before departing, the colours are much subdued by pale brown feather edges. Female is brownish apart from tail. In all plumages has *orange-red tail* which is constantly quivering. Female Redstart is similar to female Black Redstart, but is paler below (buffish grey-white) and generally warmer in tone. Agile, often catches insects in flycatcher fashion. Alarm call very similar to Willow Warbler's but usually with ticking notes at end: '**hu**eet tick-tick'. Song short, with melancholy ring, is heard as early as very first light: 'seeh tru**ee-tru**ee-**tru**ee see see seewuh…'. The verse varies all the time in its details. **SP**

Black Redstart

Black Redstart *Phoenicurus ochruros* L 14. In Britain breeds locally and uncommonly in old ruins and power stations in towns, in S Europe also very common in mountain districts. Adult male *blackish with orange-red tail* and *white wing panel*. Female similar to female Redstart but a shade darker and drabber, especially below. One-year-old male as female, but often shows some black on throat and on wing. Alarm 'weet, tk-tk-tk', with dry, treble voice. Sings, mainly at night and dawn, from elevated perch on building; a short, fast and loud phrase with a pause and *an interposed quiet crackling noise*, 'tee tee srrui chill-**chill**-chill-chill… (krshkrsh)… sreewee-wee-wee'. **RSWP**

Bluethroat

Bluethroat *Luscinia svecica* L 14. Scandinavian race *L. s. svecica* (Red-spotted Bluethroat) is common in damp willow thickets and luxuriant sub-alpine birch forest. Frequently hops on ground. Long legs. Flicks tail. In all plumages has characteristic *rust-red patches at tail base*. *Prominent supercilium*. Male's *cornflower-blue gorget* has an enamel lustre. Adult female has creamy throat framed by a varying amount of blue and black, often also some rust-red. Juvenile, as in Robin and many other Turdidae, is earth-brown with rusty-yellow spots, moulted to 1st-winter plumage before departing. Song masterful, is composed of mimicry and species-specific bell-like sounds which increase in tempo and intensity (like balalaika). Call 'trak' (like miniature Fieldfare). Alarm exactly like Wheatear's, 'heet'. Southern race *L. s. cyanecula* (White-spotted Bluethroat) breeds in reedy swamplands, turns up sporadically farther north (singing at night). In breeding plumage has small white breast spot (rarely lacking). Racial determination usually impossible in autumn. **P**

Robin

Robin *Erithacus rubecula* L 14. Breeds commonly in gardens, woods, parks and hedgerows, in boreal region typical of luxuriant coniferous forests. Fairly unobtrusive habits but not shy. Hops along on ground in long bounds, gazes in upright posture, bobs. Adults characteristically *rusty-orange on whole breast* and also up across forehead. Juveniles brown with dense yellowish-brown spotting and dark scaly pattern, moult to adult plumage by end of summer. Sings from low perch, often in cover, but also from top of trees, bushes etc. Song crystal clear, begins with very high notes, then tumbles into a lightning fast series of wildly rippling notes, checks, goes bounding off again. Autumn and winter song quieter, much more melancholy. Typical call is a series of hard, emphatic clicks, 'tic-ic-ic…', sounding like old grandfather clock being wound up. Also extremely high, thin 'tseeh' (cf. Blackbird and Penduline Tit). Nocturnal migrants from N and E Europe give a thin, weak, slightly harsh 'tsee-e'. **RSWP**

♀

♂

Redstart

♀
(*and imm.* ♂)

♂

Black Redstart

♂ *red-
spotted
form*

♂ *white-
spotted
form*

♀

Bluethroat

juv.

juv.

Robin

215

Red-flanked Bluetail

Nightingale

Thrush Nightingale

Red-flanked Bluetail *Tarsiger cyanurus* L 14. Breeds in NE Europe in luxuriant coniferous forests, often in broken ground. Has clear blue (adult male) or grey-blue uppertail-coverts and sides of tail feathers and also in all plumages *orange-tinged flanks*. Females, juveniles and first-year males (the latter hold territories) are plain like female flycatchers, especially as blue of tail is difficult to see in the field – the tail simply looks dark. *Narrow white throat patch*, framed with grey. Upperparts are olive-brown. Adult male's upperparts are dark grey-blue with clear cobalt-blue carpal areas. Most resembles Redstart in habits, but is often seen on the ground and does not quiver tail but jerks it. Shy, as a rule keeps well hidden, but most often sings from top of a spruce, often high up on steep slopes. Song is a loud, clear twitter, most closely resembling Redstart's in its thin, melancholy tone, with four or five syllables, sometimes ending in a trill. When agitated, calls with a short 'weet', and also a hard 'trak'. **V**

Nightingale *Luscinia megarhynchos* L 16.5. Breeds in thickets, in damp undergrowth in woodland and parks in S and W Europe. Rather big and long-tailed. Uniform brown with *reddish-brown tail* in all plumages. Very difficult to distinguish in the field from Thrush Nightingale, but has warmer brown (not grey-toned) back, redder tail and also *lacks obvious mottling on lower throat/breast*. Juvenile is spotted all over the body as juvenile Robin. Wary, keeps well concealed in bushes. More often heard than seen. Alarm call is a shrill, drawn-out whistle with a faint tendency towards upward inflection, '(u)eehp'. Sings from dense thickets, often at night but also by day, a loud, beautifully warbling song. In comparison with Thrush Nightingale's it is more languorous and weak, not quite so powerful; contains *typical crescendo* of soft whistled notes, 'lu lu lü lü lee leee'. The song phrases are in addition shorter, and the Thrush Nightingale's harsher, more rattling series of notes, 'zr-zr-zr-zr-…' and the like, are missing. **S**

Thrush Nightingale *Luscinia luscinia* L 16.5. Breeds fairly commonly in E Europe and the Baltic areas in leafy groves and along lakeshores in large shady copses. Has expanded towards NW in recent decades. Very like Nightingale, but can sometimes be separated by greyer back and less vivid red tail together with *diffuse grey spotting on lower throat/breast*. Juvenile looks like juvenile Nightingale (see above). Sings from concealed songpost almost throughout the day, but mostly at night. Song resembles Nightingale's in structure with series of rapidly repeated notes, but is much *more powerful and less melodic*, contains more rattling themes and characteristic very far-carrying, fast series of deep 'chok' notes. Alarm is hard, shrill, drawn-out 'eehp' notes, higher-pitched than Nightingale's, and without upward inflection, persistently repeated (rather like alarm of Collared Flycatcher). Another call is a hard rolling 'errrr'. **V**

Siberian Rubythroat *Luscinia calliope* L 15. Breeds in coniferous forest with rich undergrowth, in thickly wooded parks and similar places in the Siberian taiga. Very rare vagrant in W Europe. Male easily recognised by *ruby-red throat, white supercilium, white moustachial stripe and all-dark tail*. Female distinguished from female and juvenile Blue-throat by all-dark tail and lack of black breast band. Occasional females have a little rosy-pink on throat. Resembles Thrush Nightingale in behaviour (shy, skulking). Song powerful and melodic, *calm* and 'chatty', contains both clear and hard notes and superb mimicry. Calls: loud whistling 'ee-lu' and 'chak' ('miniature Fieldfare'). **V**

Red-flanked Bluetail

♀ (and
imm. ♂)

♂

juv.

Nightingale

Thrush Nightingale

♂

Siberian Rubythroat

♀

Wheatears (family Turdidae, genus *Oenanthe*)

Wheatears are primarily ground-dwelling birds in stony, open country. Bob and flick tail, which is black and white in characteristic pattern. The song is a short and rapid twittering and crackly phrase, often delivered by night. Catch insects, mostly on the ground, scan from perch on stones or wires, swoop down to catch the prey. Nest in hollows in the ground, among stones and boulders, or in a crevice. Clutches of 4–6 bluish eggs. The many species of wheatears are similar, and particularly females and young birds are often difficult to separate. A rough guidance is offered by the tail-pattern, and it is advisable to note the width of the trailing dark tail-band in relation to the total length of the tail.

Wheatear

Wheatear *Oenanthe oenanthe* L 14.5. Breeds commonly in open, stony country and prefers areas with sparse vegetation. In all plumages *tail has* characteristic *black and white pattern*. Male is recognised by *ash-grey back and crown*, black eye-stripe together with pure white supercilium in summer plumage. Faint rosy-buff tone on underparts of newly arrived birds fades to white in summer. Female is brownish-grey above and has poorly developed dark eye-stripe but has a distinct creamy-white stripe above the eye inclining to buff in front of eye. Juvenile has plain head, pale spotting above, dark wavy barring on breast. Greenland race *Oe. oe. leucorrhoa* is slightly larger and heavier-billed, browner above and tinged rufous below; passes through Britain, especially W coast, in good numbers in both spring and autumn. Utters a sharp whistle, 'heet', and a hard 'chak'. Sings from stone or in short song flight, frequently at night. Song consists of a short, crackly and rippling phrase at fast tempo, always with whistled 'heet' interwoven. **SP**

Isabelline Wheatear

Isabelline Wheatear *Oenanthe isabellina* L 16. Breeds in SE Europe on steppes and in semi-desert. *Large*, uniformly sandy-coloured with *paler wings* than other wheatears (visible especially in flight). In particular, lesser and median *secondary coverts and also primary-coverts are pale buff, so that brownish-black alula feathers stand out* against these. (Wheatear, which can be confusingly similar in autumn, has dark feather centres in wing-coverts, and alula does not contrast.) Tail as a rule has more black than in Wheatear. Has long and comparatively strong bill. Sexes usually alike, males on average with slightly stronger contrast, having black lores.' Supercilium whitish, especially in front of eye (cf. Wheatear). Often observed to *stand a little more upright* than is usual for Wheatear. Call a high and metallic pipe, 'cheep'. Song completely different from other wheatears' in its length – can continue for 15 sec. – its 'chattering' character and its mimicry. Series of clear 'wee-wee-wee-wee' often included. **V**

Black Wheatear

Black Wheatear *Oenanthe leucura* L 18. Breeds in SW Europe in arid, rocky mountains. Is the only wheatear in Europe that is *uniformly black*; males are jet black, females and juveniles brownish-black. All have white on rump and tail in wheatear-fashion. Also *obviously bigger* and heavier than the other wheatears. Utters a 'tshek tshek' and also shrill whistles, 'peee-e' (downwards inflected), sometimes repeated in series. Song a varied, short trill, lower-pitched than other wheatears, slightly resembling Rock Thrush's. **V**

Wheatear

♂

♀

juv.

Isabelline
Wheatear

♂

♀

Black Wheatear

Pied Wheatear

Pied Wheatear *Oenanthe pleschanka* L 14.5. Breeds in SE Europe on stony, arid slopes and erosion gullies, in west to E Bulgaria. Often perches high above ground (trees, telegraph wires), from where it flies down on to ground like a shrike to catch insects. Male has *black back which meets on sides of neck with the large black bib*. Tail pattern roughly as in Wheatear, but the *black terminal generally narrower* and can be broken up (white reaches very tip of tail) but on the other hand be more extensive on tailsides. Female extremely like female Black-eared Wheatear of eastern race, but is more *dull grey-brown* (not so warm buff-brown) *on the back*; also, the majority has a *more obvious dark throat bib* (only rarely so in Black-eared), and below the dark throat there is a tawny-coloured pectoral band (more orange-buff in Black-eared). In autumn (fresh plumage) has pale fringes to rather earth-brown body-feathers creating hint of scalloped effect. Calls include hard 'tack' and typically cracked buzzing 'brzü'. Song very like Black-eared's, short twittering phrases, sometimes with interwoven mimicry. **V**

Black-eared Wheatear

Black-eared Wheatear *Oenanthe hispanica* L 14.5 Breeds commonly in open, stony or rocky country, including maquis and vineyards, at lower level than Wheatear. Tail pattern as Pied Wheatear's. Male occurs in two colour morphs, one with black throat and one with white throat. Black-throated morph distinguished from Pied Wheatear by *buff*, not black *back*, from male Desert Wheatear by much white on tail and a pale gap between head and wing. Female distinguished from female Wheatear by no more than an ill-defined supercilium, on average *browner*, less grey *back*, often *paler lores*, more of an orange-buff pectoral band, a little more white in tail, and, on a few, grey-mottled throat. Female of eastern race (*melanoleuca*) breeding in SE Europe and Asia Minor is grey-brown on the back (not so buff-brown as western race *hispanica*) and requires great care and skills to distinguish from female Pied Wheatear (which see). Calls include hard 'tack', usually followed by whistling notes. Sings in flight (considerably more often than Wheatear) or from perch. Song like Wheatear's, though individually quite variable, sometimes mostly dry and twittering, at other times quite clear and thrush-like. **V**

Desert Wheatear

Desert Wheatear *Oenanthe deserti* L 15. Very rare visitor in autumn from breeding sites in Africa and Asia. Frequents dry, open country. Male resembles male of black-throated morph of Black-eared Wheatear, but only the rump is white – the *tail-feathers are almost wholly black* (only the innermost corners white, hard to see). Furthermore, the lesser and median wing-coverts are largely pale sand-coloured (dark in Black-eared), forming a *pale patch below shoulders*, and the black of wingbend is connected with the black of head. Greater coverts and tertials have on average broader whitish edges than in Black-eared, remaining visible longer even in worn plumage. Female resembles females of Black-eared and Pied Wheatears, lacking a pale supercilium or other strong patterns on buffish head, but all-dark pattern of tail is characteristic, and has same pale patch on lesser coverts as male (though less prominent). Usually is pale-throated, but a few have grey-mottled throat. Calls with drawn-out, shrill whistles, 'piieh', and hard clicking 'zack'. Song is a short *descending trill*, very *plaintive* in tone. **V**

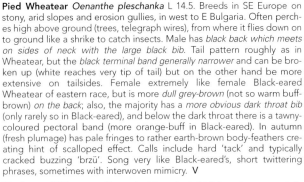

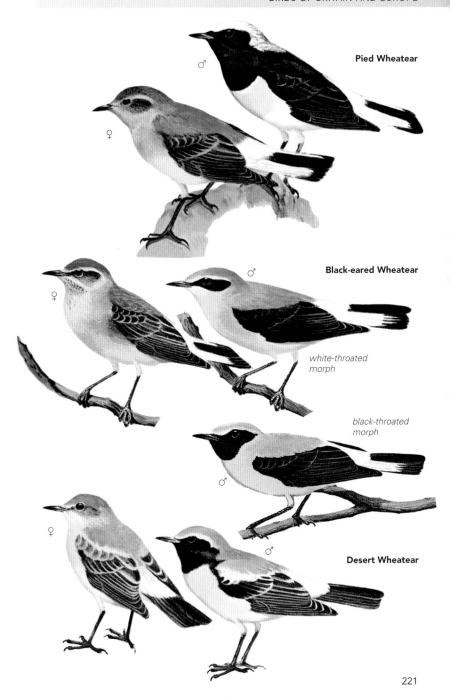

Pied Wheatear

♂

♀

Black-eared Wheatear

♀

♂

white-throated morph

black-throated morph

♂

♀

♂

Desert Wheatear

Whinchat

Stonechat

Rock Thrush

Blue Rock Thrush

Whinchat *Saxicola rubetra* L 12.5. Breeds fairly commonly on open commons, damp tussocky meadows with scattered bushes, mosses and heaths, in low vegetation along ditches. Male has prominent, *white supercilium*, and the *rusty-orange coloured throat is distinctly white-edged*; cheeks and lores very dark. White marks on wing. Female duller in colour. Distinguished in all plumages from European Stonechats by *white base to outer tail feathers* (noticeable in flight but hardly conspicuous, and can be missing in females); rump and *uppertail-coverts always streaked* brown. Perches upright, usually on top of a low bush, thistle or fence wire. Often flicks tail. Alarm call 'yu tek, yu tek-tek', in other words a soft whistle and a pure tongue-clicking. Song is short and fast, beginning and ending abruptly; a few clear notes mixed with occasional creaking sounds. Phrase variable, often contains mimicry; one variant is like Corn Bunting song. Sings mostly at night. **SP**

Stonechat *Saxicola torquatus* L 12. Breeds in central and S Europe, incl. Britain and Ireland, on heaths and grassy plains with bushes, often gorse. Usually prefers more broken country than Whinchat. Male's *all-black head* including throat is characteristic. Female has duller, brown head. Both sexes have some *white on innerwing*, male also *white on rump*. White rump usually partly streaked dark, but a few worn summer males can have a largely unstreaked white rump. *No white on tail* (exception: Stonechats from Caspian region and Iran). Siberian race (*maurus*) – rare late autumn vagrant to W Europe – has *unmarked buff-white rump*, is in fresh autumn plumage almost as pale as female Whinchat; pale supercilium, although not so prominent as in Whinchat, and pale sandy throat contrast with darker buff breast; lacks dark moustachial stripe of Whinchat. The Stonechat perches upright, appearing short-tailed and round-headed. Alarm 'weest trak trak', in other words a shrill whistle and an 'impure' tongue-clicking. Song short, has features of both Dunnock's (squeaky voice) and Whitethroat's (phrasing). **RS**

Rock Thrush *Monticola saxatilis* L 19. Breeds in S Europe in rocky and mountainous areas with or without scattered trees, usually at fairly high altitude but sometimes also at lower levels. Migratory, winters in Africa. Male in summer easily recognised by its attractive, variegated plumage. Female brownish with crescentic barring. In winter plumage male is more like female, but is recognised at close range by intimation of *blue-grey and white feathers on head and back*, respectively. In all plumages short *orange-red tail* is characteristic. Can quiver tail like Redstart. Unobtrusive, often hides among rocks. Call a short 'chak'. Sings from perch or, less often, in song flight (ends with glide), a tuneful, fluting phrase which is very variable. At a distance can recall Blackbird song, but at shorter range sounds different, is faster and more varied and embellished, lacks deep fluted notes. Not such a melancholy ring as Blue Rock Thrush's but otherwise quite like latter's. **V**

Blue Rock Thrush *Monticola solitarius* L 22. Breeds in S Europe, in particular on sheer precipices exposed to the sun and in ruins, usually at lower level than Rock Thrush. Male easily recognised by *blue plumage. Bill strikingly long*. In winter the blue becomes slightly more greyish-black. Female has crescentic barring and resembles female Rock Thrush, but has *dark grey-brown*, not rusty-brown *tail*. Some females attain bluish-grey tinge on upperparts and sides of head but invariably are mottled grey-brown on centre of throat and breast. Readily perches out in the open on a rock but usually seen only at long range, for is very shy and retiring in habits. Like Rock Thrush, disappears quickly among the rocks when disturbed. Call a hard 'tik' and a deeper 'chuk'. Song recalls Rock Thrush's, is loud and clear like latter's but more melancholy, resembles Crested Lark's song with Mistle Thrush voice. Often Blue Rock Thrush song is recognised by the 'quivering' quality of the notes. Delivered from perch as well as, more rarely, in song flight. **V**

Whinchat

juv.

♀

♂

Stonechat

♀

♂

juv.

Rock Thrush

♀

♂

Blue Rock Thrush

♀

♂

Thrushes (family Turdidae, genus *Turdus*)

The true thrushes are medium-sized birds with fairly slender though not weak bills, and rather long to medium-length tails. All are spotted in juvenile plumage, several species also when older. Seen mostly on the ground but also in trees. Feed on worms, insects and berries. Mainly nocturnal migrants. Gregarious outside the breeding season. Build open cup-shaped nests in trees or bushes. Clutches of 3–6 eggs.

Blackbird

Blackbird *Turdus merula* L 24. Breeds commonly in gardens, parks and woods (incl. deep forests). Male easily recognised by *jet-black plumage* and bright *yellow bill*. Bill darkens in autumn and early winter. Female is recognised by almost uniformly dark brown plumage. Bill mainly brown with a little yellow at base, but a few have more extensive yellow on bill. Juvenile has paler and warmer colours and has pale, narrow flecks above. Bill always dark. Feeds right out in the open on the ground. Distinguished from Starling by lack of pale spots in the plumage, by *long tail* and two-legged hops or nimble steps succeeded by dead still gaze (not waddling restless gait). Calls include: a startled series of tongue-clicking 'chak-ak-ak-ak' notes which may be heightened to an intense, shrill 'pli-pli-pli-pli-...' (before going to roost and confronted by, e.g. Tawny Owl); a deep, slowly repeated 'kok'; an extremely thin, high, 'tseeh' (similar to Robin); a slightly rolling, ringing 'srree' (also heard from night migrants). Sings from well-visible song post, mostly at dusk and dawn. The song is very tuneful and pleasing, consists of clear fluting, slowly sliding, alternating high and low notes, almost always followed by a quieter, short twitter. **RSWP**

Ring Ouzel

Ring Ouzel *Turdus torquatus* L 24. In Britain breeds fairly commonly on hilly moorland with rocky outcrops and scrubby areas, on the Continent in alpine spruce forest. *White, crescent-shaped breast band* distinguishes adult. Male not so jet-black as Blackbird, and in particular the *wings are paler*. Female is browner in tone, and the whitish crescent has brown wavy barring (but occasional individuals are confusingly like male). Juvenile is less uniform than juvenile Blackbird, e.g. has whitish-buff throat and distinctly spotted breast; is moulted by end of summer. Some 1st-winter birds confusingly all-dark, though still showing tendency to pale wings. Shy and wary. On migration associates with other thrushes. Utters a hard tongue-clicking, 'tek-tek-tek-tek' but also soft, shrill, more Fieldfare-like chatter. Song has dialectal variants but is always simple and melancholy, e.g. 'trink-trink-trink' or 'teelü-teelü-teelü', followed by a quiet twitter; resembles certain dialects of Redwing song, but pace is calm as Song Thrush's. **SP**

Black-throated Thrush *Turdus (ruficollis) atrogularis* L 23. Breeds in easternmost Europe and in Siberia in taiga, both in closed forests, in clearings and at forest edges. In winter seen in more open country and in gardens. Rare visitor to W Europe, mainly during late autumn and winter. Rather *pale* for a thrush. Male has *black throat, sides of neck and breast* with sharp border to the *unmarked white belly*. *Tail blackish, underwing orange.* Female similar, but black 'bib' of male replaced by streaks (throat) and blotches (breast) with *much white on chin and centre of upper throat*. In worn summer plumage, adult female attains broad black band across breast. First winter male similar to autumn adult female but has chin and throat densely streaked dark. First winter female more sparsely streaked on whole 'bib'. Calls e.g. squeaky Fieldfare-like 'gvieh' and hard 'chuck'. Chattering series, 'chet-chet-chet-...' also noted. **V**

Blackbird

♀

♂

juv.

♂ winter

♀ spring

Ring Ouzel

juv.

♀ 1st-winter

♂ 1st-winter

♂ 1st-winter

Black-throated Thrush

Redwing

Redwing *Turdus iliacus* L 20. Breeds commonly in open mountain birch forest in far north, less commonly at lower levels (incl. in small numbers in Scotland). In winter in fields, often loose flocks mixed with other thrushes, and in open woods. *Prominent supercilium, streaked underparts and rusty-red flanks together with underwings.* Call 'gak', alarm a persistent 'trett-trett-trett-…'; on migration an inhaled, thin, slightly harsh 'steeef', often heard on October nights, not least over towns and cities, particularly when misty. The song has many local variations. Consists of a short series of melancholy notes, usually on descending scale, followed by a low, squeaking chatter. Common variants are 'tree trü tru tro', '**chirre**-**churr**e-**chuh**ee', very fast 'til-lil-lil-lil-lil', 'tee**dye**-tee-**diue**' and more simple 'truee-traee'. A buzzing chorus is heard from flocks resting on spring migration. **RW**

Song Thrush

Song Thrush *Turdus philomelos* L 22. An extremely abundant European species. Breeds commonly in woodland, parks, gardens, hedges and in rough terrain with good cover, also in deep forest in Fenno-Scandia. Less gregarious than Redwing. *Uniformly brown upperparts,* spotted underparts together with *rufous-buff underwing* distinguish it from all other thrushes. Call a short, sharp, 'zit', also heard from night migrants. Alarm a persistent, sharp, fast 'tixtixtixtix…'. Powerful song, with fluted notes alternating with shrill, sharp notes. Many elements are repeated two to four times at calm tempo, which is very typical, e.g. 'kuklee**wee** kuklee-**wee** kuklee**wee** … kru-kru-kru … kwee-kwee … **pee**oo **pee**oo **pee**oo … chuwu-**ee** chuwu-**ee**…' and so on. Often mimics. **RSWP**

Mistle Thrush

Mistle Thrush *Turdus viscivorus* L 28. Breeds fairly commonly in woods, parks, in groves with scattered conifers. Shy and wary. In winter often on fields, usually in small numbers, associates with other thrushes. In Britain will also enter larger gardens. *Big,* heavily spotted below (belly spots rounded) with *white underwings* and uniformly *grey-brown upperparts* together with *white tips to outer tail feathers.* Has a hint of dark 'tear' below eye. Stands more upright on ground than other thrushes. Call a characteristic *dry and churring* 'zerrrrr'. Song similar to Blackbird's but has more desolate ring, the phrases are shorter and delivered at faster tempo and with shorter pauses than in Blackbird; also lacks latter's slow and considerable shifts in pitch as well as the final twittering notes. Song may run, e.g. 'tru**ee**trüwu … chu**ree**chu**ru** … chü**wü**tru … churuwütrü' and so on. Does not join in the general thrush chorus at dawn and dusk, prefers to sing on sunny mornings and afternoons, and then dominates the neighbourhood. Alarm a hard rattle, like Fieldfare's corresponding call but drier. **RW**

Fieldfare

Fieldfare *Turdus pilaris* L 25. Scarce and local breeder in Britain but common in N and E Europe, in parks and most types of forests bordering open ground. Typical of subalpine birch forest. Nests solitarily or in colonies in treetops (Merlin often nests in Fieldfare colonies in far north). Hardier than other thrushes. Common winter bird in Britain; large flocks haunt fields and plunder rowan trees and bushes carrying berries. Better flier than most other thrushes, migrates by day in large loose flocks. Flight slightly undulating. Big. *Head and rump grey, long tail dark* and *back chestnut-brown. Spotted yellowish-brown breast, pale belly. White underwings.* Call a loud frothy 'shak-shak' and a thin, slightly nasal and strained 'geeh'. Song unobtrusive, a few chattering or squeaky notes separated by brief pauses. Alternatively has an unmusical squeaky chatter, delivered without pauses for breath as bird flies among the trees. Gives hard rattle when dive-bombing crows with their droppings. **WP**

Redwing

Song Thrush

Mistle Thrush

Fieldfare

Warblers (family Sylviidae)

Small, very active, on the whole uniformly-coloured birds with slender, straight and pointed bills. In most species the sexes are alike. Song species-specific, often differing widely between closely related species. Some species are identified best by the song, or by examination of the wing formula when studied in the hand. (Relative length and shape of primaries are often conclusive.) Not gregarious. Found in dense vegetation. Feed on insects, some also on berries. Migrate at night. The warblers are divided into genera. Each genus usually includes birds with similarities in plumage and shape, in habits, and in habitat preference. The number of species treated in the book is given below after each generic name:

Grasshopper Warbler
(Locustella)

Genus *Locustella* (4 species). Tail broad and strongly rounded at the tip, faintly barred, colours brown. Live in open marshy areas with sedge, reed, bushes or other dense vegetation. Nocturnal singers. All species have a mechanical, insect-like buzzing song.

Reed Warbler
(Acrocephalus)

Genus *Acrocephalus* (8 species). All are brownish above. Head shape characteristic, with sloping forehead and long bill. Found for the most part in reeds and swamps. Song loud and repetitive, often contains harsh, jittery noises. Several species are excellent mimics.

Cetti's Warbler
(Cettia)

Genus *Cettia* (1 species). The Cetti's Warbler has a characteristically large, rounded tail; in other respects very like an *Acrocephalus*. Short, explosive, loud song phrase.

Zitting Cisticola
(Cisticola)

Genus *Cisticola* (1 species). The Zitting Cisticola is small, brown and streaked and also has a rounded tail. Frequents open ground. Odd song, delivered in peculiar song flight.

Genus *Hippolais* (6 species). Closely related to *Acrocephalus*. Grey-green or brown above, yellow or white below. Bill long with broad base, crown high with sloping forehead. Found in gardens, parks and woods. Song in a couple of species melodious and with much mimicry.

Icterine Warbler
(Hippolais)

Genus *Sylvia* (13 species). Have more richly coloured plumages than the other warblers, sexes often dissimilar. Bill short but stout, head shape rounded. Fairly shy, usually found in scrubby areas, gardens and park type country. Song melodious and warbling and sometimes thrush-like. Some species also eat berries and flower buds besides insects.

Blackcap
(Sylvia)

Genus *Phylloscopus* (12 species). Often called leaf warblers. Small and green-toned above. Short, thin bill. Yellowish-white or white below, pale supercilium and sometimes one or two pale wingbars. The different species have similar appearance but quite distinct songs. Inhabit woodlands.

Willow Warbler
(Phylloscopus)

Genus *Regulus* (2 species). Very small, green with distinctive pattern on wing. Adult birds have orange/yellow and black markings on the crown. Inhabit dense wood. Song high-pitched, not very conspicuous.

Goldcrest
(Regulus)

Savi's Warbler

Savi's Warbler *Locustella luscinioides* L 14. Breeds in S and central Europe, incl. SE Britain (rare), in thick reedbeds. *Unstreaked, reddish grey-brown above,* pale below with pale *rufous-buff breast, flanks and vent.* Undertail-coverts pale rufous-brown, in some with indistinct light tips. *Weak supercilium.* Tail long, broad and rounded. Legs dark, reddish-brown. Alarm 'pisst, pisst'. Also calls 'pex'. Song resembles Grasshopper Warbler's, but is delivered at a deeper pitch and is faster, lacks Grasshopper Warbler's reeling, ringing quality, sounds more of a *hard buzzing,* 'surrrrr…'. Can be confused with reeling of mole-cricket. Song often begins quite gently. Often sings from exposed sites in reeds or bushes, mainly at dusk and dawn but also quite often during the day. **S**

River Warbler

River Warbler *Locustella fluviatilis* L 13.5. Breeds in E Europe, rare visitor to W Europe. Lives among shrubbery and sparse alder growth on damp ground with rich undergrowth. *Unstreaked olive-brown upperparts,* pale underparts with sparse *diffuse streaking on throat and breast.* Poorly marked pale supercilium. Tail broad and rounded. Undertail-coverts are pale brown with white tips. Keeps well hidden in dense vegetation. Call harsh. Sings from bush or smallish tree. Typical *Locustella* song with quality of machine or insect, but very distinct from Grasshopper Warbler's in having *fast shuttling rhythm of sewing machine* with well separated syllables, an energetic 'zezezeze…'. Most closely resembles a low-frequency but very loud wartbiter bush-cricket *Decticus verrucivorus.* At close range a metallic background noise can be heard. Sings mostly at dusk and dawn. **V**

Grasshopper Warbler

Grasshopper Warbler *Locustella naevia* L 13. Fairly common but local in damp, open country with dense ground vegetation, prefers tussocky meadows with isolated bushes. Also occurs on drier ground such as young conifer nurseries and cornfields. Numbers vary annually. A migrant from Africa: arrives mid Apr/May, departs Aug/Sep. Plumage colours vary somewhat, but typical are heavily *streaked olive-brown upperparts, diffuse supercilium* and pale underparts, often with a few, indistinct spots on the breast. Indistinct supercilium together with heavily streaked back are best field characters. Identification, however, is best made from the song. Lives well out of sight and keeps well hidden in the dense ground vegetation. Takes flight only reluctantly when disturbed, and then flies away as short a distance as possible and very close above the grass. Call a short 'check'. Sings from low song post, mostly at dusk and dawn. The song is a fast, dry *'interminable' reeling,* like a small alarm clock with muffled clapper. Goes on for long periods without a break. The song seems to change volume as the bird turns its head. Savi's Warbler has similar song, but Savi's is shorter, deeper in tone and drier and harder. River Warbler 'chuffs'. Cf. Lanceolated Warbler's song. **S**

Lanceolated Warbler

Lanceolated Warbler *Locustella lanceolata* L 12. Breeds in easternmost Europe in dense vegetation beside swamps, marshy ground and along lakeshores, often also in open damp taiga with undergrowth and herbs. Very rare autumn and spring vagrant to NW Europe. Resembles Grasshopper Warbler, and is at times very hard to distinguish from it in the field, but is *smaller* (compact!) and *clearly streaked on entire upperparts* (streaking blacker and more distinct) and also has *heavy streaking on breast,* often some obvious streaks on flanks and sometimes also on throat. (The most heavily streaked Grasshopper Warblers are similar, but the streaks on these are predominantly found on throat; any flank streaking is diffuse, not narrow and sharp as in Lanceolated.) *Tertials are almost black with narrow, distinct pale edge* (brownish-grey with slightly broader and more diffuse edge in Grasshopper). Supercilium poorly marked. As in Grasshopper Warbler the colours may vary. Habits and call resemble Grasshopper Warbler's, and the bird keeps itself permanently well concealed in the vegetation. Extremely hard to flush. Song is quite like Grasshopper Warbler's 'svirrrrr . . .', but sounds slightly sharper, higher-pitched and more metallic, has *faint suggestion of River Warbler's 'chuffing'.* **V**

Savi's Warbler

River Warbler

Grasshopper Warbler

variation

Lanceolated Warbler

231

Reed Warbler

Reed Warbler *Acrocephalus scirpaceus* L 12.5. Breeds commonly in reeds. May also visit dense shrubbery, bushy gardens etc. Skilfully weaves a well-formed nest basket which is suspended around three or four reed stems out in the reedbed. Often exploited as foster host for young Cuckoo. *Unstreaked, brown upperparts* with *only faint suggestion of supercilium* separate it from Sedge Warbler, which occurs in similar habitats. Warmer brown above than adult Marsh Warbler; rump always rust-coloured. Note: juvenile Marsh is almost as warm brown above as Reed, can be practically inseparable even in the hand. Song, heard mostly at dawn and dusk, is not so fast and varied as Sedge Warbler's, which it otherwise resembles. The *predominantly harsh notes, repeated two or three times*, are delivered at *fussily chattering tempo*, 'trett trett trett tirri tirri trü trü . . .' etc. Like other *Acrocephalus* species, readily imitates other bird calls. **S**

Marsh Warbler

Marsh Warbler *Acrocephalus palustris* L 13. Breeds in rank, damp areas of weeds along marshy banks and river margins, preferably among nettles and meadowsweet, also in drier growth of reeds with interspersed weeds. Only local in Britain. Very similar to Reed Warbler but has a slightly shorter bill, and adult is more *greenish brown* above. Also has slightly paler legs than Reed, and tips of primaries have fine pale edges. Difference between juveniles of Marsh and Reed is even slighter, both being warm brown above. As a rule, different habitat (Marsh Warbler is rarely seen in dense reed-beds) will do as a guide. Song (best in early morning but also heard at late dawn and night) is elaborate and quite different from Reed's. *Tempo is typically varied*; sometimes song slows right down, sticks on mimicking calls which are repeated (cf. Blyth's Reed Warbler), but then the bird identifies itself by an *acceleration* and an *explosion of masterful mimicry* (often including calls of Blue Tit, Magpie, Swallow, Chaffinch and Blackbird) interwoven among series of quite unmelodic, dry trilling calls (also these are imitations, but of tropical species). A characteristic harsh 'zi-**cheh** zi-**cheh**' is an obligatory element. **S**

Blyth's Reed Warbler

Blyth's Reed Warbler *Acrocephalus dumetorum* L 12.5. Breeds in NE Europe (incl. S Finland) in damp woodland edges and glades with dense thickets. Very rare summer visitor to NW Europe. Resembles Marsh Warbler, but rump slightly more rusty in tone and wings more rounded, primary projection thus being shorter. Bill slightly longer than Marsh's. Supercilium more marked than in Marsh, reaching just beyond eye. Alula and tertials pale brown, make wing even more lacking in contrast than in other *Acrocephalus* species. Also on average darker, greyer feet than its relatives (but still have brown element). The song, heard mainly at night, is given from rather a high perch. Brilliant mimic. Song, however, very distinct from Marsh's in its *calm tempo* and fact that *every phrase is repeated* five or six, sometimes even ten times. Many individuals have a characteristic phrase ('the stairs'), a clear, rising, 'lo leu **lee**a', with long note breaks. Nearly always *interposed between the phrases is a typical tongue-clicking* 'chek chek'. **V**

Paddyfield Warbler

Paddyfield Warbler *Acrocephalus agricola* L 12. Breeds at W Black Sea and on steppes east of Caspian Sea, in reedbeds where replaces Reed Warbler. Rare vagrant to W Europe. Resembles Reed Warbler, but is *smaller* and has *more distinct supercilium, emphasised by dark crown edge*. Bill also *smaller* (comparatively short), usually dark upper mandible, pale yellow-brown lower, dark tip. Upperparts vary somewhat but usually are paler than in Reed Warbler. Can be confused with Booted Warbler (p.238) but has more rounded tail without white, and rump always has a rusty-brown tone. Wings strongly rounded. The call is described as 'tschik'. Song is forcedly fast and 'jumpy', rich in mimicry, fairly low, immediately recalls Marsh Warbler's but lacks latter's tempo changes, trills and the crescendo as well as typical 'zi-**cheh**'. **V**

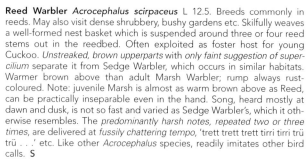

Reed Warbler

adult

Marsh Warbler

juv.

adult

Blyth's Reed Warbler

juv.

adult

Paddyfield Warbler

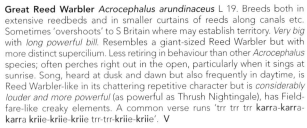

Great Reed Warbler *Acrocephalus arundinaceus* L 19. Breeds both in extensive reedbeds and in smaller curtains of reeds along canals etc. Sometimes 'overshoots' to S Britain where may establish territory. *Very big with long powerful bill*. Resembles a giant-sized Reed Warbler but with more distinct supercilium. Less retiring in behaviour than other *Acrocephalus* species; often perches right out in the open, particularly when it sings at sunrise. Song, heard at dusk and dawn but also frequently in daytime, is Reed Warbler-like in its chattering repetitive character but is *considerably louder and more powerful* (as powerful as Thrush Nightingale), has Field-fare-like creaky elements. A common verse runs 'trr trr trr **karra-karra-karra** kriie-kriie-kriie trr-trr-**kriie-kriie**'. **V**

Great Reed Warbler

Sedge Warbler *Acrocephalus schoenobaenus* L 12.5. Breeds commonly in reedbeds in swamps and in other dense vegetation along lakeshores and riverbanks, sparingly in the far north in osier beds. Upperparts are streaked, albeit quite lightly (in worn plumage the back is fairly uniform grey-brown). *Supercilium distinct* and long, *buff-white or dirty-white.* Resembles Moustached and Aquatic Warblers (cf. these). Rump unstreaked yellowish rusty-brown. Juvenile can have faintly streaked breast and a faint crown-stripe, though never so obvious as in the considerably paler and yellower Aquatic Warbler. Call 'tsek'. Alarm a very hard rattle 'trrr'. Often sings in pitch-dark night (Reed Warbler prefers dawn and dusk). In daytime often performs a short vertical song flight. Song *varied, full of mimicry and harsh, jittery calls.* It resembles Reed Warbler's but *the tempo is more hurried* and more varied, giving a feverish and hectic quality. (Remember the basic rule: Sedge Warbler Spirited Singer, Reed Warbler Relaxed.) Often recognised by rapid accelerating crescendo of excited notes that turns into tuneful whistles, e.g. 'zreezree trett zreezreezree trett, zreezreezree **pseet** trutrutru-peerrrrrrrrrrrr-urrrrrr wee-wee-wee **lulu-lu** zitri zitri…' etc. **S**

Sedge Warbler

Aquatic Warbler *Acrocephalus paludicola* L 12.5. Rare and local breeder in central and E Europe in wet meadows with low growth of *Carex* species. Winters in W Africa; migration route westerly, passes through France. Like Sedge Warbler, but has *paler yellow-brown plumage* and a *pale longitudinal stripe on the crown.* (Note that juvenile Sedge can have rather a pale central line.) Distinct yellowish-white supercilium. *Heavily streaked back* with *pale, buff longitudinal stripes on the mantle,* and strongly streaked rump. Adult often has a few distinct narrow streaks on breast, whereas the young are unstreaked below or has just a few grey spots. Keeps well concealed in the vegetation. Song resembles a 'sleepy' Sedge Warbler's. Most audible is a dry 'trrrr', which opens most phrases, and which is sometimes repeated for long periods – it may sound like Sedge Warbler alarm. Normal stanzas consist of just a few motifs, the flat trills and various whistled notes, 'trrrr … veeveeveevee … cherrr kyee-kyee-kyee…' . Sometimes, particularly in short song flights, the stanzas become slightly longer. **V**

Aquatic Warbler

Moustached Warbler *Acrocephalus melanopogon* L 12.5. Breeds in S Europe in reedbeds and swamps with dense vegetation, often stands of bulrushes. Resembles Sedge Warbler, but has more *reddish-brown back, darker crown and ear-coverts* which contrast with *purer white super-cilium* and white throat. Flanks and sides of breast tinged rufous. Supercilium is *broad and abruptly square-ended.* Sometimes gently cocks its tail. Often sings from well visible song post. Song resembles Reed Warbler's, but it is slightly softer and livelier and is recognised sooner or later by a *series of drawn-out, rising, Nightingale-like whistling notes,* 'we we wee wih'. Alarm 'trrrt'.

Moustached Warbler

Great Reed Warbler

Sedge Warbler

adult

juv.

adult

juv.

Aquatic Warbler

Moustached Warbler

235

Cetti's Warbler *Cettia cetti* L 14. Breeds in S and W Europe in dense and low vegetation by ditches, watercourses and marshes. Has in recent years expanded northwest and now has an established breeding stock in England. Keeps well concealed and is always difficult to see. Note *unstreaked, dark reddish-brown upperparts*, greyish-white underparts and *narrow whitish supercilium* (a bit Wren-like). *Tail broad and rounded.* Often jerks tail downwards or sideways. Wings short and rounded. Bill slender and pointed. Sexes alike. Heard more often than seen. Has an explosive tongue-clicking/cracking 'pex'. Alarm a furious 'pexexex…', rather similar to Wren. Song begins suddenly and ends abruptly, *is very loud and explosive*, has a fleeting stop halfway through, often sounds 'tsi-tsi-chuut! chuti-chuti-chuti!'. ('What's my name? **cetti cetti cetti**!') Sings from well-concealed perch. **R**

Ceti's Warbler

Zitting Cisticola (Fan-tailed Warbler) *Cisticola juncidis* L 10. Breeds in S Europe in open country, on plains with tall grass, in cornfields etc. Recognised by *small size*, clear streaking on crown and upperparts together with *very short, rounded tail* with black and whitish tip. Very unobtrusive, keeps well hidden except for extensive, bounding song flights at c. 10 m height. These take place even in middle of day at height of summer. Song consists of series of slowly repeated, penetrating 'zrip' notes, one 'zrip' with each bounce in the undulating flight. Alarm a quick 'chic-chic-chic-…'. **V**

Zitting Cisticola

Icterine Warbler *Hippolais icterina* L 13. Breeds rather commonly in eastern half of Europe in deciduous woods with undergrowth, dense parks and larger gardens. Greenish-grey upperparts and usually *clear pale yellow underparts, including belly*, distinguish it from other warblers except very similar Melodious, which replaces it in SW Europe. Pale edges to secondaries form *pale panel on folded wing* in fresh plumage. Face fairly plain, lacks dark lores and eye-stripe. Long-winged. *Legs blue-grey.* Perches more upright than *Phylloscopus* species and is not so restless and active. Often raises crown feathers when excited. In alarm gives series of hoarse, nasal, tongue-clicking notes, 'tettettet-tett…', heated and with rather uneven rhythm. The clicking note is included in call, a short, musical 'tete-**lu**eet'. Song very varied and pleasing, full of masterful mimicry. It also contains grating notes, and a *nasal and creaky* '**gee**a' is peculiar to the species. Some individuals repeat the verses more pedantically than most, and may even throw in a 'tett' (risk then of confusion with Blyth's Reed Warbler). Sings from crown of tree. Usually does not sing at night. **P**

Icterine Warbler

Melodious Warbler *Hippolais polyglotta* L 13. Breeds in SW Europe in open deciduous wood with rich undergrowth, parks and gardens and also riparian scrub. Much resembles Icterine Warbler, but is a shade *yellower below* and *browner grey above*, has shorter wings, often *brown legs* and *less prominent pale edges to secondaries*. (Yellow underparts not always striking in the field; can in fact be taken for Marsh Warbler.) *Call a House Sparrow-like chatter*, 'krrrrrr'. Song faster and more chattering, does not contain so many imitative notes as Icterine's. The rattling call is interwoven into the song, but Icterine's nasal '**gee**a' is always lacking. Sometimes less inspired, slow and repetitive song can be heard, creating resemblance to Isabelline Warbler song (and 'melodious' becomes a less apt name). **P**

Melodious Warbler

Cetti's Warbler

Zitting Cisticola

adult, spring

juv.

Icterine Warbler

adult

Melodious Warbler

juv.

Isabelline Warbler

Olivaceous Warbler

Olive-tree Warbler

Booted Warbler

Isabelline Warbler (Western Olivaceous Warbler) *Hippolais opaca* L 14. Breeds in S and SE Iberia and in NW Africa in open woodland, maquis, gardens and orchards. Winters in W Africa. Closely related to Olivaceous Warbler, and until recently regarded as a race of this. Extremely rare vagrant north of its range. Differs from Olivaceous Warbler on being slightly larger and *longer-tailed*, and on having *stronger legs and bill*, latter being both longer and broader (slightly convex sides in lateral view). Also, it *keeps its tail still* when moving (does not pump it downwards). Upperparts pale sandy-brown, or isabelline-coloured, without obvious paler wing panel. Legs greyish-pink, lower mandible of bill pinkish-yellow. Song can recall Reed Warbler's, is perhaps a trifle slower than that of Olivaceous and more articulate, varied and pleasing. Song often starts with a few repeated calls, 'chek'. Other calls include slurred trills, 'chrrrt', like in Olivaceous.

Olivaceous Warbler *Hippolais pallida* L 13. Breeds in SE Europe in a variety of wooded habitats, tall scrub, riverine forests, parks and gardens. Spends winter in E Africa. *Upperparts basically greyish with faint olivaceous tinge* when fresh, underparts pale, and wings and tail darker grey. In fresh plumage there can be a *hint of a paler secondary panel*, and folded secondaries can appear white-tipped. Bill rather long and pointed, *lower mandible unmarked pinkish-yellow*. Legs greyish with very faint pinkish tinge. Has peculiar habit of *repeatedly pumping tail downwards* when moving in canopy, *often accompanied by a clicking 'chek' call*. Song a rather monotonous phrase which mixes lower scratchy notes with high-pitched nasal, squeaky ones in a cyclic pattern, going up and down the scale in a repetitive way. Tempo is not fast but 'energetic', and articulation is rather 'blurred' with notes running into each other. Apart from the mentioned tongue-clicking 'chek' call has conversational subdued 'chrrrt' which may also be loud and drawn-out, then more like Melodious Warbler. **V**

Olive-tree Warbler *Hippolais olivetorum* L 15.5. Breeds in open oakwoods, in orchards, maquis, or open grassy country with scattered dense bushes and trees. Largest member of *Hippolais*, slim with long wings and tail. *Primary projection long, equals tertial length*. Tail and *exposed primaries very dark* above, and centre of secondary edges prominently edged white, forming *whitish patch or panel* (unless plumage too worn). *Very long, pointed bill, lower mandible pinkish-orange*. A pale line above lores, and whitish eye-ring, but *no real supercilium*. Moves confidently, with no hurry. Often wavers tail to sides and down, as if loosely-fixed, but can also pump it downwards in Olivaceous-fashion. In flight like small thrush, often glides with outstretched wings before landing. Song very deep and raucous (Great Reed Warbler-voice), staccato-like (jerky pace), cyclically repeated. Can be likened to song of Olivaceous Warbler played at too slow speed. Call a deep, hard, tongue-clicking 'chack!'. Has a quarrelling nasal but hard 'kerrekekekek' when alarmed, almost like 'giant Blue Tit'. **V**

Booted Warbler *Hippolais caligata* L 11.5. Breeds in NE Europe in bushes and scrub, often in overgrowing, neglected pasture land or on drier meadows. Main distribution on wide steppe in Central Asia. Westward-expansion in recent decades; has bred Finland. Rare autumn vagrant in W Europe. Often feeds low in vegetation or on ground, yet difficult to see, keeps well hidden. Smaller than Olivaceous Warbler, which it otherwise recalls, with shorter wings, tail and bill (almost *Phylloscopus* impression). '*Milky-tea*' coloured above, whitish below with *buff-brown tinge on breast and flanks*. *Pale supercilium* reaches a bit behind eye. Tail-sides diffusely edged paler, tail rather square-cut. Bill has pale lower mandible with darker tip. Legs dull pinkish, toes greyer. Does not pump tail downwards, but often *twitches both wings and tail very fast* in Willow Warbler-fashion when moving. Song a very *fast irregular warbling* without clear structure; *starts subdued but picks up pitch and strength*. Call a tongue-clicking 'chrek', slightly compound, with 'r' in, like soft Stonechat call. **V**

238

Isabelline Warbler

Olivaceous Warbler

Olive-tree Warbler

Booted Warbler

Barred Warbler

Western Orphean Warbler

Eastern Orphean Warbler

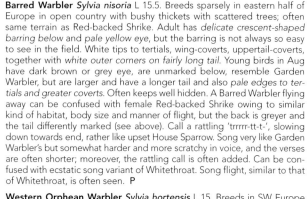

Barred Warbler *Sylvia nisoria* L 15.5. Breeds sparsely in eastern half of Europe in open country with bushy thickets with scattered trees; often same terrain as Red-backed Shrike. Adult has *delicate crescent-shaped barring below* and *pale yellow eye*, but the barring is not always so easy to see in the field. White tips to tertials, wing-coverts, uppertail-coverts, together with *white outer corners on fairly long tail*. Young birds in Aug have dark brown or grey eye, are unmarked below, resemble Garden Warbler, but are larger and have a longer tail and also *pale edges to tertials and greater coverts*. Often keeps well hidden. A Barred Warbler flying away can be confused with female Red-backed Shrike owing to similar kind of habitat, body size and manner of flight, but the back is greyer and the tail differently marked (see above). Call a rattling 'trrrrr-tt-t-', slowing down towards end, rather like upset House Sparrow. Song very like Garden Warbler's but somewhat harder and more scratchy in voice, and the verses are often shorter; moreover, the rattling call is often added. Can be confused with ecstatic song variant of Whitethroat. Song flight, similar to that of Whitethroat, is often seen. **P**

Western Orphean Warbler *Sylvia hortensis* L 15. Breeds in SW Europe including Italy in open woodland, groves and parks. *Large with dark hood, or dark ear-coverts*, narrow *whitish iris* (except in young and some females, which have darker eye) and white on outer tail-feathers. Upperparts dull dark brown, underparts cream-white with *pinkish-brown tinge on flanks and vent*. Similarly coloured Sardinian Warbler is considerably smaller and has red orbital ring in adults. Immature birds, and some females, can resemble a large Lesser Whitethroat, but size difference is 15–20%, and bill of Western Orphean is considerably stronger and tail longer. Song is loud, deep-voiced and rather uncomplicated, one or two motifs repeated a few times, e.g. '**tee**ro **tee**ro **tee**ro' (can recall Ring Ouzel). Calls include a clicking 'chak' and rolling, hard 'terrr'. **V**

Eastern Orphean Warbler *Sylvia crassirostris* (Not illustrated.) L 15.5. Breeds in SE Europe, from Slovenia and eastward, in deciduous woods on mountain slopes, in orchards, groves and parks. Closely related to Western Orphean Warbler, and until recently regarded as a race of it. Very similar in appearance as well, and without song or close observation vagrants often inseparable. Differs from Western Orphean in following ways: *underparts whiter* (although lower flanks and vent still tinged buff-grey); *undertail-coverts have dark centres* (difficult to see, though); black hood of adult male slightly more contrasting, and rest of upperparts very slightly greyer; bill on average a little longer, and base of lower mandible a little paler grey (more contrasting), but much overlap. *Song is much more elaborate and pleasing*, recalls both Nightingale and Blackbird, e.g. 'tru tru tru sheevu sheevu sheevu, yo-yo-yo-bru-trih'. Calls as in Western. **V**

Blackcap

Blackcap *Sylvia atricapilla* L 14. Abundant European breeder, dominating the chorus in many wooded habitats like lush deciduous forests, parks and larger gardens, in the Mediterranean area also in more tall-grown maquis, in N Europe also in mature spruce forest with glades. Medium-sized grey warbler. *Black cap which stops above the eye* distinguishes the male, and *reddish-brown cap the female* and immatures. (In autumn, young males may have brown cap with mixture of black feathers.) Easier to catch sight of than the shy and elusive Garden Warbler, but nonetheless usually keeps well concealed. Hardy, capable of enduring quite severe cold. Often eats berries, is not dependent on insect diet. Usually sings from well-concealed song post. Song resembles Garden Warbler's in general tone, but verses are shorter, and the initial rippling babble *turns towards the end into a few characteristic, clear and powerful, fluted notes* with a melancholy ring. Rarely a subsong full of mimicry. Call is a hard clicking 'tek', like Lesser Whitethroat but louder. Alarm is a series of hard, loud, hurried tongue-clicking notes, 'tek-tek-tek-tek-…'. **SWP**

Barred Warbler

1st autumn

♂
adult

Western Orphean Warbler

♂ *1st winter*

♀ *breeding*

♂ *breeding*

♂

Blackcap

♀

♀

241

Garden Warbler

Garden Warbler *Sylvia borin* L 14. Breeds commonly in open woodland with rich undergrowth and also in larger thickly wooded gardens and parks. *Olive-tinged brownish-grey upperparts* and *greyish-white underparts*, legs grey-brown. Lacks particular distinguishing features. Most easily confused with Olivaceous, Booted, Marsh and juvenile Barred Warblers. Note rounded head profile, the *short and rather heavy grey bill* and also almost total lack of supercilium. Often slightly grey-toned on sides of nape/hindneck. Keeps well hidden in the foliage even when singing. Song ripples forth in a charming babble, like Blackcap's, but the verses are longer and lack latter's clear final notes. Voice is rather deep, and since the song lacks a clear tune it has been likened to the sound of a rippling creak. Alarm a hoarse note with a slightly nasal clucking quality, 'interminably' repeated, 'tshek-tshek-tshek-…'. **SP**

Whitethroat

Whitethroat *Sylvia communis* L 14. Breeds commonly in scrub and bushy areas (common e.g. on heaths with blackberry and low juniper), in cultivated country wherever hedges and shrubbery are present. Fond of rank plant vegetation, thrives in more open habitats than Lesser Whitethroat. *White throat*, buff-toned breast (pink tinge in adult male), brown-grey head (greyest in adult male), *pale reddish-brown wings* and long tail with pale outer feathers are characteristic. Best distinguished from Lesser Whitethroat by *paler legs* and reddish-brown wings. Iris pale grey-brown to ochrous. A hint of a white eye-ring in most. Very active, constantly on the move in bushes and scrub. Call a harsh 'whed whed whed'. Alarm a harsh, drawn-out 'chairr'. Song is delivered from a bush top or telegraph wire and is a rapid and quite short, jerky verse, harsh in tone; a common rhythm runs **chuck**-a-ro-**che**, **chuck**-a-ro'. Now and again the Whitethroat flies up a few metres and performs a protracted, ecstatically chattering flight song of rather general *Sylvia* character. **SP**

Lesser Whitethroat

Lesser Whitethroat *Sylvia curruca* L 13.5. Breeds quite commonly in dense bushy thickets, in hedges in gardens and in young pine groves. Resembles Whitethroat, but is a shade smaller with shorter tail, grey-brown upperparts and *dark grey ear-coverts*. Wings uniformly grey-brown, lacking rufous edges of Whitethroat. Most of *underparts very pale*, almost pure white, centre of breast without buff tone as in, e.g. female Subalpine Warbler, which can be a confusion risk in the Mediterranean countries. *Iris grey, legs dark grey*. Keeps well concealed. Call a short tongue-clicking 'tett'. Numerous in SE Europe (Asia Minor) on migration, where has a further call, a Blue Tit-like hoarse scolding 'chay-de-de-de'. Song, delivered usually from concealed songpost, consists of two parts: first a brief, muffled chatter, which turns into a fast, loud, rattling trill, 'tellellellellellellell'. **SP**

Spectacled Warbler

Spectacled Warbler *Sylvia conspicillata* L 12.5. Breeds in SW Europe in open, dry localities with scrub. Looks like a Whitethroat but is *smaller*, and slimmer and has *finer bill* (dark, with yellowish-white base). Adult male has darker grey forecrown shading to *blackish on lores and around eye*, a *more obvious white eye-ring* and darker pink breast (a little greyish on lower throat). Female very similar. Distinguished from Subalpine Warbler female by *reddish-brown wings*, but a few are tricky, being less vividly rufous. Call a very dry, clear, rattlesnake-rattle, 'zerrrrr'. Song a typical *Sylvia* chatter; fast tempo, quite short phrases, *high-pitched voice*; the phrases usually open with one or a few notes of Crested Lark clarity. Sings from perch in full view or during song flight. **V**

Garden Warbler

Whitethroat

♂

♀

Lesser Whitethroat

♂

♀

♂

Spectacled Warbler

Subalpine Warbler

Rüppell's Warbler

Sardinian Warbler

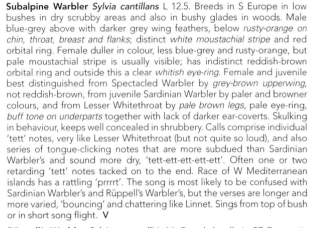

Subalpine Warbler *Sylvia cantillans* L 12.5. Breeds in S Europe in low bushes in dry scrubby areas and also in bushy glades in woods. Male blue-grey above with darker grey wing feathers, below *rusty-orange on chin, throat, breast and flanks*; distinct *white moustachial stripe* and red orbital ring. Female duller in colour, less blue-grey and rusty-orange, but pale moustachial stripe is usually visible; has indistinct reddish-brown orbital ring and outside this a clear *whitish eye-ring*. Female and juvenile best distinguished from Spectacled Warbler by *grey-brown upperwing*, not reddish-brown, from juvenile Sardinian Warbler by paler and browner colours, and from Lesser Whitethroat by *pale brown legs*, pale eye-ring, *buff tone on underparts* together with lack of darker ear-coverts. Skulking in behaviour, keeps well concealed in shrubbery. Calls comprise individual 'tett' notes, very like Lesser Whitethroat (but not quite so loud), and also series of tongue-clicking notes that are more subdued than Sardinian Warbler's and sound more dry, 'tett-ett-ett-ett-ett'. Often one or two retarding 'tett' notes tacked on to the end. Race of W Mediterranean islands has a rattling 'prrrrt'. The song is most likely to be confused with Sardinian Warbler's and Rüppell's Warbler's, but the verses are longer and more varied, 'bouncing' and chattering like Linnet. Sings from top of bush or in short song flight. **V**

Rüppell's Warbler *Sylvia rueppelli* L 14. Breeds locally in SE Europe in thorny scrub in mainly rocky areas. Male easily recognised by *black hood and throat* separated by striking *white moustachial stripe*, and grey upperparts. Female has more of a dark grey hood, and often has *diffusely dark-spotted throat* separated from dark head by thin white line; looks all-grey above with *light edges to very dark tertials* and a few greater coverts. Orbital ring and *legs reddish-brown* in both sexes. Call a sparrow-like rattle, like Barred Warbler's but weaker and no slowing down as in that species, and also a series of hard tongue-clicking notes, harder than corresponding call of Subalpine Warbler, not so much like a 'mechanical rattle' as in Sardinian Warbler. Song very like Sardinian's but not quite so loud; more 'chugging', like a pulsating rattle, e.g. 'prr-trr-trr prr-trr-trr see-tree-wee-prr' and similar. Song flight with slow-motion flight like Greenfinch and occasional glides on upwards-angled wings. **V**

Sardinian Warbler *Sylvia melanocephala* L 13. Breeds commonly in S Europe in scrub and bushes in open or rocky country, but also in woods with undergrowth. Characteristic of the male are *jet-black hood* with well visible *reddish-brown orbital ring*. Female and juvenile more difficult to identify, but note fairly *dark upperparts* and *grey-brown flanks* together with *long* rounded *tail*, which is blackish with white sides (often striking in flight). Underparts look sullied, contrasting with white throat. Female also has rather prominent reddish-brown orbital ring, juvenile a more indistinct brown one. The two Orphean Warblers are much bigger, have stronger, longer bill and lack red orbital ring. Calls are an explosive chattering mechanical rattle, 'churrrr, trit-trit-trit-trit-trit' or slower 'terit terit terit terit', and also, when agitated, series of more grinding tongue-clicking notes or occasional loud, hard 'tsek'. Song a typical *Sylvia* chatter. Pace speedy, length of verses varying but normally 2–5 sec. Composed of hard 'trr-trr' call-like notes intermingled with utterly short whistles. Song flight like Whitethroat's. **V**

♂

Subalpine Warbler

♂ 1st winter

♀

♂

Rüppell's Warbler

juv.

juv.

♀

Sardinian Warbler

♂

Dartford Warbler

Dartford Warbler *Sylvia undata* L 13. Breeds in SW Europe (incl. in S England) in dry, bushy localities, on heath with scrubland, often among gorse, broom and thornbushes. Mainly resident. In northern part of the range the birds are very hard pressed during severe winters and in consequence numbers there fluctuate considerably. *Long-tailed, very dark grey-brown above, dark red-toned below, throat diffusely white-spotted.* Female's brown colour considerably duller, can appear entirely grey at swift glance. Darker than any other warbler except Marmora's and Balearic, which have grey, not red-brown, breast. Often clearly demarcated white belly. Bill base yellowish. Often holds tail cocked and frequently flicks it. Flight low and weak with characteristic jerky tail movements. Usually keeps well concealed in the vegetation. In winter sometimes seen in small flocks which rove about like tits. Calls are a harsh, drawn-out and slightly inflected 'chaihrr-er', repeated at a grinding tempo when agitated, and also a 'tak', which may be repeated in a rattling, fast series. The song is rather rugged and hard-voiced. Normal phrase length 2–3 sec. Can be heard singing throughout the year. Song flight often performed. **RS**

Marmora's Warbler

Marmora's Warbler *Sylvia sarda* L 13. Breeds locally on islands in C Mediterranean region (Corsica, Sardinia, Elba, Pantelleria) in dry, scrubby terrain, often in rocky country. Partly migratory, some winter in N Africa. Resembles Dartford Warbler, but has *grey*, not deep red-brown *underparts* (although reddish element in Dartford surprisingly difficult to see in some lights). Female is slightly browner above and a little paler below, juvenile still browner and paler. Adult male told from Dartford also on *lack of well-defined white belly* and *lack of white-spotted throat.* Long, narrow tail and dark colours distinguish it from all other warblers except very similar Balearic Warbler (which see). Habits like Dartford's, keeps mostly hidden in cover. Song recalls Sardinian's and Spectacled's, is a fast twittering warble usually ending with a clear trill or flourish. Call is a throaty 'chrek', almost like Stonechat (by some even likened to Dipper). **V**

Balearic Warbler *Sylvia balearica* (Not illustrated.) L 12. Breeds only on Mallorca and Ibiza. Resident. Found in maquis and garrigue in rocky country. *Smaller* and *longer-tailed* than extremely similar Marmora's Warbler, from which it has recently been separated, recognised as a species of its own. Male on average *slightly paler grey below* with *whitish throat* (instead of grey). Legs are more vividly orange-tinged, also in juvenile (juv Marmora's has duller brown legs). Apart from on locality, best told on voice. Song a short and fast outburst like in Marmora's, but drier and more mechanically rattling up and down, a bit like Sardinian Warbler alarm. Call a nasal, muffled 'tsreck'.

Bush Robins (family Turdidae, genus *Cercotrichas*)
Bush Robins is a genus of smaller thrushes which are closely related to nightingales (*Luscinia*) and the common Robin (*Erithacus*). They feed much on the ground, have strong legs and long tails, which they often fold up above their backs or fan. Frequents open country with bushes. For a presentation of the other thrushes, see pp. 214, 218 and 224.

Rufous Bush Robin

Rufous Bush Robin *Cercotrichas galactotes* L 15.5. Sparse breeder in S Europe in open, dry and bushy habitats, in vineyards, hedgerows and gardens, preferably in prickly pear hedges. Arrives late in spring, not until May. Two clearly separated races in Europe: race *galactotes* in Iberia has rufous-brown crown and back, whereas race *syriacus* in SE Europe is paler and more grey-brown. Characteristic of both races is the *long rufous tail* with striking *black and white markings at the tip. Pale supercilium* and dark eye-stripe. Often behaves quite boldly, perching right out in the open with tail raised and fanned, demonstrating its prominent pattern. Often seen on the ground. Sings frequently, usually from top of bush, sometimes in pipit-like parachuting song flight. Song can recall both Robin and Song Thrush through its high-pitched notes and reciting delivery. Can last minute-long in a fairly even rhythm. Call a sharp, clicking 'tak'. **V**

Dartford Warbler

♂

juv.

Marmora's Warbler

juv.

♂

Rufous Bush Robin

*eastern
(grey-brown back)*

*western
(rufous-tinged back)*

Willow Warbler

Chiffchaff

Iberian Chiffchaff

Wood Warbler

Willow Warbler *Phylloscopus trochilus* L 11.5 Very common summer visitor, in deciduous and mixed wood from lowland copse to highest sub-alpine birch, requires simply a group of trees with undergrowth. Even grey-green in tone above. *Breast has touch of yellow.* Juvenile has fairly strong yellow underparts. Populations in north and northeast average more grey-brown. Similar to Chiffchaff but usually less brown in tone, has more distinct supercilium and eye-stripe, as a rule *paler leg colour* (but can have quite dark brown legs). Is very active and restless like other *Phylloscopus* warblers; scampers about in tree foliage in search of insects. Call and alarm call a weak, soft 'hooeet'. Song is slightly melancholy but pleasing; starts with a few high, clear notes, drops down the scale and slows down, but regains speed only to drop again and die away in soft languorous notes. **SP**

Chiffchaff *Phylloscopus collybita* L 11. Common breeder in parks, large gardens and all kinds of woods with undergrowth, also in mature spruce forests in Fenno-Scandia. Small numbers winter in Britain. Resembles Willow Warbler but is generally a trifle smaller, has more *grey-brown tinge in olive of upperparts*, and head and breast are a little duskier, *supercilium often shorter, bill is finer and darker*, and legs are dark grey-brown or blackish (a few exceptions, but good rule of thumb). Unlike Willow Warbler *repeatedly pumps its tail down* when moving in canopy. Siberian race *tristis* rare but regular vagrant to W Europe, told on having supercilium, throat and breast buff-brown or ochrous-tinged instead of yellow; only yellow in plumage is wing-bend and underwing. Upperparts grey-brown with slight green tinge on back, rump and edges of wings and tail. European Chiffchaffs call like Willow Warbler but not so clearly disyllabic, has stress more on second syllable, is a little stronger, not so weak and soft, 'hüeet'. Siberian *tristis* calls with monosyllabic, straight, piping call, 'eehp'. Song in Europe rather plain, a long series of single notes uttered at moderate pace, 'chiff chiff cheff chiff chaff chiff chiff…'. When agitated often intersperse shivering 'perre perre' between series. **RSP**

Iberian Chiffchaff *Phylloscopus ibericus* L 11.5. Breeds in W Iberia; also in extreme SW France. Closely related to Chiffchaff, which it formerly was treated as a race of, and which it hybridises with to some extent in Basque region. Migrant, wintering in tropical W Africa. Mainly found in elevated open deciduous woods. Like a mixture of Chiffchaff and Willow Warbler: *greenish above, pale yellow and white below* with virtually no brown or grey tinges. Supercilium usually quite yellow, especially in front of eye. Bill a little stronger on average than Chiffchaff's, with more pale brown on base. Legs brown, neither very dark nor pale. *Dips tail down repeatedly* when moving in canopy. *Song best means of identification*, contains three elements (order can be altered, and all three are not always used in each phrase), a Chiffchaff-like part, a stammering part and a whistling part, e.g. 'chief chief chief chief tr-tr-tr-tr swee swee swee'. Call is also diagnostic, whistling and downwards-inflected, full 'seeuu'. **V**

Wood Warbler *Phylloscopus sibilatrix* L 12. Fairly common in woods with tall trees with no or scant undergrowth ('pillared hall'), therefore typical of mature beech woods, but also found in mixed forests and large parks. *Vividly bright green above, lemon yellow on throat and breast* and pure white on belly. *Supercilium yellow and prominent*, enhanced by *dark eye-stripe. Long and pointed wings* make tail appear short. Song characteristic and easy to recognise, an accelerating series of sharp 'zip' notes ending in metallic trill ('coin spinning on marble slab'), 'zip… zip… zip, zip, zip zip zip-zip-zip-**zip-zwirrrrrr**'. Short horizontal song flights frequent, low between trees, giving 'zip' notes in flight and finishing trill upon landing. Often switches to totally different alternative song, a melancholy piping which increases in intensity, 'dyuh dyuh **dyuh-dyuh-dyuh**'. Call a sharp 'zip', alarm single 'dyuh'. **S**

Willow Warbler

*adult
North Europe*

juv.

*adult
North Europe*

adult spring

Chiffchaff

*Siberian
race* tristis

Iberian Chiffchaff

Wood Warbler

249

Western Bonelli's
Warbler

Eastern Bonelli's Warbler

Arctic Warbler

Greenish Warbler

Western Bonelli's Warbler *Phylloscopus bonelli* L 11. Breeds in C and SW Europe in wooded habitats of many kinds (oak, pine, etc.) and at varying altitudes (sea level to over 1500 m). Pale olive-brown upperparts with *bright yellowish-green rump and edges to wings and tail.* (Paler rump at times less obvious in juveniles.) *Tertials dark with contrasting white tips* and yellow-green edges. Underparts whitish with only faint buffish tinge to sides of breast. Moderately marked pale supercilium. *Unbroken white eye-ring* distinguish it from all other members of the genus except Eastern Bonelli's Warbler. Ear-coverts pale buffish-grey, rendering the bird a rather *bland-faced look.* Legs usually dark brown-grey. Usually keeps well concealed in the foliage, but now and then hovers in the open at the tip of a branch. Call a loud and sharply whistling, distinctly disyllabic 'hü-**eef**'. Song a brief repetition of one note in a loud and clear trill, almost laughing in tone (rather like a part of Wood Warbler song), 'svi-vi-vi-vi-vi-vi-vi'. **V**

Eastern Bonelli's Warbler *Phylloscopus orientalis* L 11.5. Breeds in SE Europe (mainly Greece and Bulgaria) in closed woods, often of oak. Closely related to Western Bonelli's Warbler and until recently treated as a race of it. Very similar to Western Bonelli's, and autumn birds often inseparable on plumage. In spring, Eastern Bonelli's tend to have more abraded tertials and greater coverts, these feathers being greyish forming a hint of a paler wing panel. Crown to back tend to be a trifle duller grey-brown (less tinged olive), but this is difficult to use on single birds. Best told on voice. Call is a monosyllabic chirping 'chip', much like a begging young House Sparrow. Song very similar to that of Western Bonelli's but a little faster, drier, more mechanical and insect-like. **V**

Arctic Warbler *Phylloscopus borealis* L 12.5. Rare and local breeder in far north in birch woods on mountain slopes, usually near watercourses. Arrives extremely late due to long migration (winters in SE Asia, over 12,000 km away!). Large, slim, rather big head and strong bill, distinguished by one *whitish wingbar, long and distinct yellowish-white supercilium* (stopping short of nostrils; cf. Greenish Warbler) and *very dark and distinct eye-stripe* (reaching base of bill). Legs usually rather pale pinkish-brown. Upperparts rather greenish-tinged, *underparts* whitish, generally *with diffuse oily-grey mottling or streaking on breast and flanks.* Call peculiar for a leaf warbler, a sharp, short, penetrating 'dzre!' (like Dipper). Song a fast reeling trill, 'sre-sre-sre-sre-sre-…', rather like Cirl Bunting song. **V**

Greenish Warbler *Phylloscopus trochiloides* L 11. Breeds in E Europe in tall, mature woods with undergrowth. Arrives late, at end of May; a few overshoot to W Europe. Shares with Arctic Warbler *one whitish wingbar* (though this usually thinner and shorter in Greenish) but told on *smaller general size,* more *grey-tinged upperparts, less dark and distinct eye-stripe* (particularly in front of eye) and *supercilia* which *reach past nostrils* (or meet over base of bill). Call a sharp disyllabic 'zee-lee', rather like Pied Wagtail. Song a high-pitched, sharp and hurried twittering (like agitated Pied Wagtail-song); at times a Wren-like song variant is heard. **V**

Dusky Warbler *Phylloscopus fuscatus* L 11. Rare autumn vagrant from E Asia. In short, a dark brown bird with buff-white underparts (tinged grey), lacking wingbar and which calls with a *tongue-clicking* 'tack' (like Lesser Whitethroat). *Supercilium long,* all whitish or tinged rufous at rear. Recalls Radde's Warbler but has *finer bill* and *thinner and darker brown legs,* and *supercilium is more distinctly outlined in front of eye,* and is whiter there. V

Radde's Warbler *Phylloscopus schwarzi* L 12. Rare autumn vagrant from E Asia. For separation from similar Dusky Warbler, note faint olive tinge in brown of upperparts and *stronger ochrous-buff tinge on flanks and vent.* *Supercilium is generally slightly diffusely outlined in its foremost part* (against fore-crown) and is *tinged buff* (Dusky is white). *Legs and bill stronger,* legs also paler pink-brown. Call a nasal, slurred 'chrep'. **V**

Western Bonelli's Warbler

Eastern Bonelli's Warbler

spring

Arctic Warbler

Greenish Warbler

Dusky Warbler

Radde's Warbler

Yellow-browed Warbler *Phylloscopus inornatus* L 10. Breeds in extreme NE Europe and in Siberia, in taiga. Rare but regular autumn visitor to NW Europe, mainly in late Sep–Oct. *Very small* with greenish upperparts, off-white underparts, pale yellow-white, *long supercilium, two distinct white wingbars*, the lower one broad, and *broad white tips to tertials*. Lower wingbar enhanced by *black bases to secondaries*. Sometimes a faint suggestion of a paler central crown-stripe at rear of crown (but nothing like in Pallas's). Ear-coverts and cheeks off-white finely blotched grey-green (no buff tinge). Very active, dashing around in foliage when feeding. Often associates with other leaf warblers and with tits. Frequently calls, a characteristic high-pitched, drawn-out whistle, 'tsu-**weest**', a little like Coal Tit but higher-pitched, longer and invariably upwards-inflected. Song is very high-pitched and thin, does not carry far, can be described as mixture of Hazel Grouse song and Goldcrest calls. **P**

Pallas's Warbler *Phylloscopus proregulus* L 9.5. Rare but regular autumn visitor to NW Europe, mainly in late Sep–Oct, from breeding area in Siberia. Although a breeder in closed taiga with much coniferous trees, appears on migration in Europe in deciduous woods, gardens, willow stands, shrubbery and thickets. *Very small* and resembles Goldcrest, but has conspicuous, *bright yellow and distinctly outlined rump patch* (rarely whitish). This is particularly striking in flight or when hovering. Like Yellow-browed Warbler has *double white wingbar* (lower one quite broad) and *white-tipped tertials*, but differs on rump patch and *narrow yellowish central crown-stripe. Sides of crown* and *eye-stripe very dark olive-grey. Forepart of long and prominent supercilium chrome yellow*, and whole forehead often tinged yellow. Associates with tits and Goldcrests on migration. Active, moves much, hovers, even climbs upside down like a small tit. Rather silent on migration. Call is upwards inflected 'twooeet', a little like Chiffchaff but more squeaky and nasal. **V**

Goldcrest

Goldcrest *Regulus regulus* L 9. Breeds commonly in spruce wood, also in other coniferous and even mixed woods, incl. larger gardens. Goldcrests' high, thin calls are commonly heard from high up in the trees in many mature spruce forests. To see them is harder; they usually dodge about high up, skip and climb quickly along the branches, flutter momentarily on the outside of the tip of a branch, etc. The Goldcrest is Europe's smallest bird. It has a *plain grey-green plumage* but is adorned with a *broad yellow crown-stripe, edged black*. The *male has* in addition a touch of *orange in the yellow*. Juvenile lacks the bright head pattern. Outside the breeding period Goldcrests roam around in loose groups, often with tits, and when such flocks leave the cover of a coniferous forest, as on migration, they prove to be fearless and easily approached as they perch in bushes and taller herbs in open country. Call is a high-pitched, shrill note, generally repeated three times, 'sree-sree-sree' (difficult to hear for elderly people). Alarm a piercing 'tseet'. Song is also high and thin, a few notes repeated in cyclic pattern and ending in a flourish, 'sesu**si**-sesu**si**-sesu**si**-sesu**si**-sesu**si**-sesu**si**-susisee**swirrr**'. **RWP**

Firecrest

Firecrest *Regulus ignicapilla* L 9. Breeds in C and S Europe in deciduous, coniferous and mixed woodland, and in parks. Resembles Goldcrest but distinguished from that species in all plumages by a *distinct white stripe above the eye*, and a *black stripe through it. Shoulders/sides of neck are rather bright bronzy-green* (in Goldcrest dull grey-green like the back), which is characteristic. Most easily told from Pallas's Warbler by lack of sharply demarcated pale rump patch. Habits as Goldcrest's. Call even sharper and shriller than Goldcrest's, differing in that it sounds more intense and that pitch and intensity raises slightly at the end, 'zezeze**zi**'. Song is a monotonous series of high notes on same pitch (no cyclic pattern like in Goldcrest!) that accelerate and rise slightly; it is not so well articulated as Goldcrest's, and the tempo is a shade faster. **RWP**

Yellow-browed Warbler

Pallas's Warbler

adult ♂

adult ♀

Goldcrest

adult ♂

Firecrest

adult ♀

Flycatchers (family Muscicapidae)

Small birds which catch insects in flight and build nests in cavities or on ledges. Lay 4–9 eggs.

Pied Flycatcher

Pied Flycatcher *Ficedula hypoleuca* L 13. Breeds commonly in woods with at least some deciduous element, in parks and gardens but also in remote forests. Readily uses nestboxes. Not quite such a specialist at taking flying insects as Spotted Flycatcher. Flicks wings (often one higher than other) and cocks tail when perched on look-out. Some males (especially in E Europe, occasionally further west) are browner, though always have *white forehead patch* (female never has). Male is easy but female very difficult to distinguish in the field from corresponding sexes of Collared Flycatcher (see that species). Song quite powerful, sprightly and rhythmic, e.g. 'see tsee**vree** tsee**vree** tsee**vree** yu lee tsee**plee** tsee**plee** tsee**plee** tsee**plee**'. Alarm a persistent, short metallic 'pik, pik, pik…'. Also gives quiet tongue-clicking 'tett'. **SP**

Collared Flycatcher

Collared Flycatcher *Ficedula albicollis* L 13. Breeds fairly commonly in deciduous wood and gardens in C and SE Europe. Male told from male Pied by broad *white neck collar, large white forehead patch*, noticeably more white on wings and *greyish-white rump*. Many females separable in the field from female Pied under favourable circumstances by greyer upperparts and more obvious white on folded wing (*pure white also on bases of primaries*; none or only a little buffish-white on inner primaries on female Pied). Song surprisingly unlike Pied's, consists of drawn-out squeaky notes, as if pumped and squeezed out. Alarm persistently repeated, loud 'eehlp' notes. Also gives quiet tongue-clicking call. **V**

Semi-collared Flycatcher

Semi-collared Flycatcher *Ficedula semitorquata* L 13. Breeds sparsely in SE Europe (Bulgaria, Greece). Closely related to both Pied and Collared Flycatchers. Male resembles Pied but has more white on wing, including *white-tipped median coverts* and *bigger white patch at base of primaries*. Also, has *more white on tail* and slightly larger white patch on forehead. Some males have visibly *more white on side of neck* than Pied (are semi-collared), others not. Females are very similar to Collared, are greyish-tinged brown above. Call usually a straight piping note, 'tüüp', vaguely recalling Siberian Chiffchaff. Song like a mixture of Pied and Collared, contains both rhythmical and strained notes. Weak, does not carry far.

Red-breasted Flycatcher

Red-breasted Flycatcher *Ficedula parva* L 11.5. Fairly common but local in luxuriant, shady, often slightly damp parts of woodland, deciduous or mixed. In Britain scarce migrant. Nests in tree crevices. Behaves a lot like a leaf warbler, moving actively in the foilage. Cocks tail (characteristic). *White patches on tail base.* Adult male has *rusty-orange throat*. Female nondescript buffish-white on chin/breast but shares male's typical tail pattern. First-summer males (both sing and breed) have female-like plumage. Song begins rhythmically like Pied's, ends with characteristic series of descending (not dying away) notes, 'sree … sree, sree, seewüt seewüt seewüt seewüt wüt wüt wiü wiü wiü wew'. Call a dry rolling 'serrrt', weaker than Wren. A clear '**tee**lu' (alarm, agitation) is often heard, as is a Lesser Whitethroat-like 'tek'. **P**

Spotted Flycatcher

Spotted Flycatcher *Muscicapa striata* L 14. Common in open woods, woodland edges, parks and gardens. Like the other flycatchers a true summer visitor and long-distance migrant, wintering in C Africa. Nests in recesses, often in inconventional sites (flower pots, windowsills, etc). Perches on look-out on jutting branch, launches out in short sally and snatches some small flying insect. Upright posture when perched, flicks wings. Plumage *plain brownish-grey* with streaking on breast and (diagnostically) *on forehead*. Sexes alike. Call a sharp 'zreet', alarm 'isst-tec'. Song extremely simple, consists of 3 or 4 call-like notes. **SP**

Collared Flycatcher

Pied Flycatcher

♂

♀

Collared Flycatcher

♂

♀

Semi-collared Flycatcher

♂

Red-breasted Flycatcher

♂
adult

Spotted Flycatcher

♀ *(and imm. ♂)*

255

Tits (family Paridae)

Tits are small, short-billed and agile birds. Often quite fearless. The sexes are alike, and the juveniles rather like the adults. Normally sedentary, but several species move south or west in some years. Outside breeding season often seen in mixed flocks, known as roving tit flocks, which may also contain Nuthatches, Treecreepers and Goldcrests. They often find their way to bird tables, particularly in winter. Hole nesters, readily use nest boxes. The clutches vary between 5 and 16 eggs. Eggs are white with red speckles.

Marsh Tit

Willow Tit

Sombre Tit

Siberian Tit

Marsh Tit *Parus palustris* L 12. Breeds commonly in deciduous and mixed woods and seems particularly fond of dense leafy thickets and neglected gardens. Nests in natural hole in tree or, occasionally, nest box. Does not join roving flocks as readily as Great and Blue Tits; very sedentary; often seen in pairs. Zealous hoarder of seeds, often seen low down in the undergrowth. Resembles Willow Tit, but *wings uniformly coloured without whitish panel,* bib usually smaller, cheeks not pure white. Juvenile cannot be safely distinguished from juvenile Willow Tit by appearance. Best distinguished from Willow Tit by calls. Typical are a short, explosive 'pi**chay**' and a clear and full 'cheeü', and also a hoarse, slightly Blue Tit-like, excited series, 'ziche dedededede'. The song has several patterns but is always a rapid series of full and loud notes. Common variants are 'chüp-chüp-chüp-...' (recalling Greenfinch) and '**tee**ta-**tee**ta-**tee**ta-...' ('saw-sharpening', like Coal and Great Tits). **R**

Willow Tit *Parus montanus* L 12. Breeds commonly in coniferous and mixed forest, also mountainous areas and subalpine birch forest, in Britain also in damp birch and alder woods without conifers. Usually excavates own nest hole in rotten tree stumps. In winter joins roving tit flocks. Distinguished most easily from Marsh Tit by calls (see above), and by *whitish panel along wing* formed by pale edges to secondaries and usually somewhat larger bib. British race *P. m. kleinschmidti* has buffish flanks, especially noticeable outside breeding season. Fenno-Scandian and N Russian race *P. m. borealis* is slightly paler, with purer white cheeks. Juveniles cannot be reliably distinguished from juvenile Marsh by appearance. Willow Tit's most characteristic call is 'tee-tee **cheh cheh**', the two final sounds strongly stressed, drawn-out and harsh. The song is a series of pensive and well-articulated 'tiu tiu tiu...' (recalling Wood Warbler), less commonly a straight 'teeh teeh teeh...' (the same individual may alternate between the two). Sometimes gives an odd, short alternative song phrase, rapid, cheerful and chuckling. **R**

Sombre Tit *Parus lugubris* L 13.5. Breeds in SE Europe in deciduous wood (often oak), also in mountainous areas. Big, has heavy bill, gives Great Tit-like impression. Plumage appears shabby. *Very large bib. Cap is dull brownish-black* (browner in female). Not gregarious, and more shy than other tits. One call is like Long-tailed Tit's, 'zreeh-zreeh-zreeh' with sharp, grating tone; another a very House Sparrow-like chattering series, 'cher-r-r-r'. The song recalls Marsh Tit song, but the voice is gruffer and more grating, the tempo slower: 'chii**ev**-chii**ev**-chii**ev**-...'.

Siberian Tit *Parus cinctus* L 13. Breeds sparsely in the northernmost coniferous forests, reaches right up to the subalpine birch forest. Recognised by *rusty-yellow flanks* (less obvious in summer), dull brown cap and *large black bib.* Plumage more 'bushy' than in other tits (save Azure Tit). Commonest call is a fast 'tee-tee chay chay', in which the hoarse final syllables are not so straight and drawn-out as in corresponding call of Willow Tit. The song is a thin and fast purring 'chee-**ürrr** chee-**ürrr** chee-**ürrr** chee-**ürrr** chee-**ürrr**'. Besides this, a Marsh Tit-like 'che che che che che...' and a cheerful, short, chuckling phrase 'see see di**twuy**' occur.

Marsh Tit

Willow Tit

British race

Scandinavian race

Sombre Tit

Siberian Tit

Crested Tit

Blue Tit

Azure Tit

Coal Tit

Crested Tit *Parus cristatus* L 12. Breeds in coniferous forest, mainly pine, in Britain confined to Caledonian forest of Scotland. Extremely sedentary. Nests in hole in decayed trunk, capable of excavating own nest hole. Easily recognised by well-visible *crest. Black and white head markings*, in all plumages ages and sexes alike. Sometimes associates with Coal Tit, but spends more of its time near the ground. Call a characteristic, short purring or 'bubbling' trill, 'burrurr**itt**' (somewhat recalling Snow Bunting call), and a thin '**seeh**-lili'. Song: '**seeh**-burrurr**itt**-**seeh**-burrurr**itt**-...', at a fast rate. **R**

Blue Tit *Parus caeruleus* L 12. Breeds commonly in deciduous wood (also mixed wood), parks and gardens. Outside breeding season often visits reedbeds. *Yellow underparts* as in Great Tit but has *bright blue cap* enclosed with white and is also distinctly *smaller*. Wings and tail blue. Juveniles in summer have yellow cheeks and greyish-green cap. Searches branches and trunks in typically clambering fashion, less often seen on ground. Behaviour in winter at bird table self-assured, often drives away the larger Great Tit. Has many calls, but most often heard is a clear 'seeseedu' and a quarrelsome churr 'cherrrrrr-errr-errr-ett'. The song consists of two thin, drawn-out notes followed by a crystal-clear trill, 'zeeh zeeh sirrrrr'. A song variant with two short phrases in rapid succession, 'ziziserr zizi-serr', is sometimes heard. **RW**

Azure Tit *Parus cyanus* L 13. Breeds in E Europe in deciduous and mixed woods, especially around lakes and rivers. Very rare visitor to C and W Europe. In winter often frequents reed-beds. Very white-looking. Plumage 'fluffy'. Tail rather long. *White head with black eye-stripe* and collar around nape, very *broad white wingbar, much white on tail*. Hybrids with Blue Tit occur, are more inclined to migrate, are often very similar to Azure Tit, but usually have a faint blue cap (Azure Tit has white crown) and less white on wings and tail. All calls resemble those of Blue Tit, including a quarrelling 'kerrr-ek-ek-ek' and the high-pitched, clear song, 'siih siih tetetetetete'.

Coal Tit *Parus ater* L 11. Basically a bird of coniferous forest, but in Britain also fairly common in gardens and in deciduous wood. Mass eruptions in N Europe when spruce crop fails. Outside breeding season associates with other tits. Black head with white cheeks and *white nape patch* together with faint buff tone on underparts are characteristic. Rather short-tailed. Common calls are thin, clear, very melancholy '**teeh**-e, tü'. Also less characteristic, Goldcrest-thin squeaks. The song is a fast and nimble 'sipi-**tee**-sipi**tee**-sipi**tee**-...' (totally characteristic) or a 'saw-sharpening' 'sitchew-sitchew-sitchew-...' **RW**

Great Tit *Parus major* L 14. Breeds commonly in all kinds of woodland, in parks and gardens. In some autumns sets off southwards from northern areas, when many may reach Britain (especially east coasts). Size, *glossy black crown*, white cheeks and *black band down centre of yellow underparts* make it easy to identify. Band wider and blacker in male than in female, particularly obvious on belly (see fig. at left). Young in summer have yellowish cheeks without black lower border. Often seen in mixed flocks with other tits, size then separates out the Great Tit at first glance. Often feeds in low bushes and also on ground. Calls innumerable and usually more powerful than other tits', e.g. rather Chaffinch-like 'ping-ping', a slightly 'wondering' and melancholy 'tee tü tüh' (autumn call) and also confident and rapid 'see-**yutti**-**yutti**'. Song very characteristic, a penetrating 'saw-sharpening' series, e.g. '**tee**-ta **tee**-ta **tee**-ta...' or trisyllabic '**tee**-tee-**tü tee**-tee-**tu tee**-tee-**tü**...', the last variant like Mozart's (from The Marriage of Figaro) 'Say goodbye butterfly to merriment' without the final note. **RW**

Great Tit

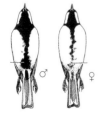

♂   ♀

Crested Tit
from below

Crested Tit

Blue Tit

Azure Tit

Coal Tit

juv.

Great Tit

259

Long-tailed Tit (family Aegithalidae)

The Long-tailed Tit is closely related to the true tits. Apart from long tail is distinguished by the architecture of its nest. Clutches of 8–12 red-spotted white eggs.

Long-tailed Tit

Long-tailed Tit, Spain

Long-tailed Tit *Aegithalos caudatus* L 16. Breeds fairly commonly in deciduous and mixed wood, often with hazel or thick bushes. Nest skilfully built, with dome covered with lichens and sited in bushes or tree forks. In winter may be seen in company with other tits but always keeps together in small flocks, even within a roving tit flock. *Very long tail* (in most of range) characteristic. British and Irish race *A. c. rosaceus* has *white head with broad black stripe above eye*, strong pinkish-buff tone on scapulars, rump and vent, and whitish wing panels. Juvenile has dark cheeks, little pink and shorter tail. Adults of northern/eastern race *A. c. caudatus* have all-white head, whiter wings, look much whiter. Intermediate forms occur in N central Europe. Spanish Long-tailed Tits are very swarthy, have stripy sides of head, wine-red tone on flanks and almost all-black back; tail is comparatively short. Resembles other tits in habits, but rarely comes to bird-tables. Calls are a dry churring 'tserr' (recalling Wren), clicking 'tett' and also a high, shrill 3-syllable 'sreeh-sreeh-sreeh'. Does not normally sing to defend territory. A silvery Blue Tit-like descending trill, sometimes heard from flocks, is not a song but alarm (against e.g. Sparrowhawk). **R**

Bearded Reedling (family Timaliidae)

The Bearded Reedling resembles the tits in most respects, but belongs nevertheless to the babbler and laughing thrush family. Clutches of 5–7 eggs are laid in open nests low down in reeds.

Bearded Reedling

Bearded Reedling *Panurus biarmicus* L 16.5 Scattered distribution in C and S Europe in larger reedbeds. Increasing and spreading in NW Europe. In summer feeds on insects, in winter on reed seeds. Easily recognised by *buff and cinnamon-brown colours* and *very long tail*. Male has distinct *black drooping moustache* and black undertail-coverts, which are absent in female and juvenile. Juvenile told from female by black lores (lacking in female), middle of back and sides of tail. Very active, clambers around in reeds with hopping movements, often at lowermost levels. Flight weak with very rapid wingbeats, usually low over reeds. In autumn, however, seen working its way high up into the sky in dense flocks, usually only to dive headlong back (still in closed formation) but sometimes to migrate. Outside breeding season seen almost exclusively in flocks. Main calls are twanging, lively 'pching' and 'pchew', very characteristic (but can be imitated expertly by Reed Warbler!). Also, a fine spinning 'chiürrrrr' is heard. Song twittering. **RW**

Penduline Tit (family Remizidae)

The Penduline Tit is a small bird resembling a true tit, with a thin, pointed bill. Sexes similar. Nests in finely woven, domed nest with entrance provided by tunnel. Clutch of 6–8 eggs.

Penduline Tit

Penduline Tit *Remiz pendulinus* L 11. Breeds mainly in S and E Europe. Locally common in shoreline thickets and shrubbery in or near reeds. Currently expanding northwestwards (several recent British records). Builds a curious pouch-shaped nest suspended from the very end of a twig of a tree (often willows or birch), close to water. A quite small bird, smaller than Blue Tit. *Black band* or 'mask' *on 'face'* distinguishes adult. Male has bright chestnut-brown back, female more yellow-brown back. Juvenile lacks the black band on the face, lacks the black band on the face, has evenly sandy head. Behaviour much like that of true tits. Call a very thin, drawn-out 'zeeeeu', as fine as Red-throated Pipit's call but dropping slightly at the end like Robin's thin peep. (If Robin can be said to sound sharply 'alert', Penduline Tit sounds if anything 'good-natured' and 'dreamy'.) Song quiet and heard only for brief moments, rather Coal Tit-like, 'zeeu-seewut zeeu-seewut zeeu-seewut' and other variants. **V**

Long-tailed Tit

southern form

northern form

nest

Bearded Reedling

♂

♀

juv.

juv.

nest

nest

♂

Penduline Tit

261

Nuthatches (family Sittidae)

Nuthatches have strong feet, large heads, short tails. Climb trees or cliffs. Pick insects out of crevices with the long, straight bill. Very agile and even move downwards on a tree trunk head first, which treecreepers and woodpeckers cannot do. This feat is accomplished by one foot being held higher up on the trunk so that the bird is almost suspended from this foot. Outside breeding season often associate with tits. Calls loud and characteristic. Flight jerky and undulating. Sexes similar. Hole nesters. Clutches of 5–7 white eggs with red spots.

Nuthatch

Nuthatch *Sitta europaea* L 14. Breeds fairly commonly in older deciduous woods, in parks and gardens. Regular visitor at bird tables during winter months. Nests in hollow trees and often reduces size of entrance hole by plastering it with mud. Has a profile all of its own, with long, pointed bill, short tail and crouched posture. *Climbs head first down trunks. Back blue-grey, long black streak through eye.* Colour of underparts in adult varies from white in *S. e. europaea* (Fenno-Scandia, N Russia) to pale reddish-brown in *S. e. caesia* (W Europe). Intermediate forms occur. Male's flanks vividly chestnut-red, female's duller rusty-brown. Does not drum like the woodpeckers, but often hammers at nuts and seeds of various kinds to get at the contents. Has several characteristic, *very loud calls.* Common call a sharp, emphatic 'seet, seet', alarm/scolding call an excited 'twett-twett-twett, twett…'; in calmer mood scolds with a 'chwit, chwit, chwit…'. Song is loud, has tone like a musical whistle, may be drawling, 'ueeh ueeh ueeh…' (alternatively 'weeu weeu weeu…') or fast, 'wiwiwiwiwi…'. A further variant is often heard, a rapid 'jujuju jujuju...'. **R**

Corsican Nuthatch

Corsican Nuthatch *Sitta whiteheadi* L 12. Breeds exclusively on Corsica in montane pinewoods, where it is resident. Looks like a small Nuthatch, but note the distinctive markings on the head with *black crown/nape* and clear, *white line above black eye-stripe.* Female and juvenile darker in colours, but markings the same. Habits as Nuthatch's, but rather shyer. Capable of excavating own nest hole in tree. Has hoarse, scolding call, 'chay-chay-chay-…'. Song consists of clear notes in very fast trills, 'hididi-dididi…', or series of more drawn-out notes, 'dew-dew-dew-…'.

Krüper's Nuthatch

Krüper's Nuthatch *Sitta krueperi* L 12. Turkish species. Breeds in Europe only on Greek island of Lesbos, in montane pinewood. Closely related to Corsican Nuthatch and accordingly small. Glossy black crown patch (does not extend to nape), black eye-stripe and blue-grey back like Corsican Nuthatch, but overall impression of face paler, and is easily recognised by *reddish-brown patch on breast.* Calls in essence very like Corsican Nuthatch's. Include Greenfinch-like 'dyuee' and hoarse scolding 'chay-chay…' and (in flight) a Brambling-like 'jek'. Song loud, monotonous, fast 'didadidadida…' or variations on one tone 'dididi…'.

Rock Nuthatch

Rock Nuthatch *Sitta neumayer* L 15. Breeds in SE Europe on rocky slopes and mountains with scattered bushes. Looks like a *big, pale* Nuthatch, but lacks latter's white spots on tail and has proportionately *longer bill.* Resembles Nuthatch in movements and posture. Climbs cliffs instead of trees. Nests on cliff faces, on which voluminous nests are plastered with mud. Very 'talkative' with loud and high-pitched calls. Song very loud, and echoing between mountainsides, long series of decelerating, clear notes that often drop deeply in pitch (creating remote resemblance with Woodlark song structure). Both sexes sing.

Nuthatch

Britain and W Europe

Scandinavia

Corsican Nuthatch

Krüper's Nuthatch

Rock Nuthatch

Wallcreeper (family Tichodromadidae)

The Wallcreeper is a close relative of the nuthatches, but has a long, curved bill and behaves like the treecreepers. Builds bulky nest in cleft in a cliff. The 4–5 white eggs have red spots.

Wallcreeper

Wallcreeper *Tichodroma muraria* L 16. Uncommon and local in occurrence in high mountains up to the snow line. Seems to be easier to find in its eastern range of distribution. Nests in the Alps usually at between 1000 m and 2500 m altitude, but nests have been found down to 350 m above sea level. In winter moves further down, and can then be seen on church towers, walls of castles and in quarries. Longer movements are very rare (though several have even reached Britain). Appearance odd, unmistakable. *Bill long and faintly curved.* Shape and pattern of *wings* very striking: *broad and round with large, red panels and white spots.* Upperparts grey, throat/breast black in summer and pale grey in winter. Female, however, has merely a black patch on throat in summer. Flight characteristically fluttering, recalling butterfly. Climbs cliffs in search of insects, all the time flickering its wings so that the red flashes. On the other hand it does not use its tail as a support like a treecreeper. Call thin and piping but loud, a whistling 'tiuh' and variations. Song consists partly of vaguely twittering series, and partly of (totally diagnostic) peculiarly strained, drawn-out notes with glissando 'tu…ruuee…zeeeeeu' repeated in the same way several times in succession. **V**

Treecreepers (family Certhiidae)

Treecreepers are small, short-legged, brown-speckled birds with thin, curved bills. Climb trees in search of food in crevices of bark. Usually seen singly. Very active. Build nest in clefts or hollows (often behind loose bark). The 5–7 eggs are white with red spots.

Treecreeper

Treecreeper *Certhia familiaris* L 13. Breeds fairly commonly in older woodland, parks and gardens, both coniferous and deciduous wood. In S Europe where it occurs within the same area as Short-toed Treecreeper, it prefers coniferous woods at higher altitudes. Typically climbs trunks in spirals, starting at the foot of the tree. Easily distinguished from other birds except Short-toed Treecreeper. Often has *more distinct pale supercilia* than latter, especially in front of eye above lores; can even meet narrowly over base of bill. Usually lacks the pale brown colour on the flanks and has *slightly paler, more contrasting plumage* and more of a red tone on rump. Bill is on average a shade shorter than Short-toed's. Call a repeated, very high, thin and sharp but yet rolling 'srrree, srrree…'. Also gives pure, plain 'teeeh' (quite pronounced, rather Coal Tit-like, may be repeated in a series, though not rapidly as in Short-toed Treecreeper). Song is a sharp and thin, rather low-voiced, clear ditty which accelerates, *falls in pitch* and *ends with a short, melodic flourish*, the phrase typically has quick, abrupt finish and Blue Tit-clear voice. **R**

Short-toed Treecreeper

Short-toed Treecreeper *Certhia brachydactyla* L 13. Breeds in C and S Europe in older deciduous woodland, parks and gardens, usually at lower altitudes than Treecreeper and this particularly so in the southern part of the range (but can commonly be found on lower hills in cedar forests, etc.). Often seen climbing upside down along horizontal branches. Very similar to Treecreeper, but *is a shade darker* and less rusty-toned above and also has *pale brown tone on flanks*. Has subtly shorter and more curved hindclaw, whereas bill averages slightly longer than Treecreeper's. Call a powerful, clear and Coal Tit-like 'tüüt', repeated at well-spaced intervals or in very typical 'dripping' series (crystal clear, penetrating voice), which accelerate to fast trotting rhythm and fall in pitch. In addition a very Treecreeper-like 'srree' may be heard. The song has a more plaintive tone (Coal Tit-like) than Treecreeper's, is a short clear phrase with a slightly *jolting rhythm*, 'tü te toh etititt', *final notes quicker and slightly rising in pitch*. **V**

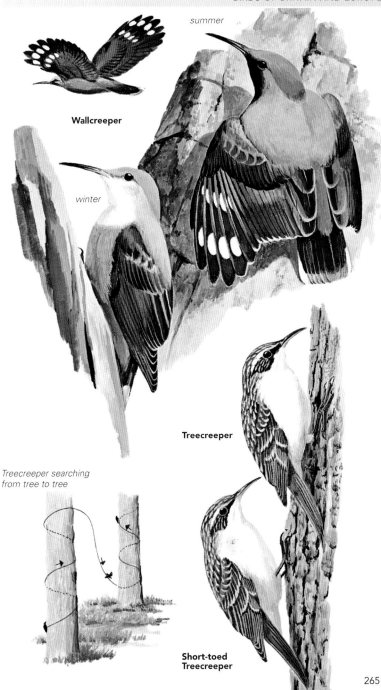

summer

Wallcreeper

winter

Treecreeper

*Treecreeper searching
from tree to tree*

**Short-toed
Treecreeper**

Shrikes (family Laniidae)

Shrikes feed on insects or small birds, which are sometimes speared on thorns or barbed wire. Often seen perched alone on bush tops or on telegraph wires in open country. 4–6 eggs.

Red-backed Shrike

Red-backed Shrike *Lanius collurio*. L 18. Breeds in open country. Now very rare in Britain. Male has *rufous-brown back, ash-grey crown/nape, black-and-white tail*, white throat and *pink-buff breast*. Female and juvenile are brown above (juvenile heavily cross-barred) and cream-white below, densely barred dark on breast and flanks. Odd females look more like males but has dark brown tail with very little white on edges, and are invariably somewhat barred beneath. Call (also used as territorial signal) is a short, hoarse 'cheve'. Agitated series of hoarse 'che-it' often heard. Sometimes gives a warbling sub-song, pleasant and full of mimicry. **RSP**

Masked Shrike

Masked Shrike *Lanius nubicus* L 18. Breeds in SE Europe in semi-open country. Smallest and *slimmest* European shrike. Note *black* (male) or *dark grey* (female) *upperparts* and *long, narrow, black-and-white tail*. Large white shoulder patches and *wing patches* (latter conspicuous in flight). Juvenile is *barred and greyish*, lacks yellowish-brown tone below, best identified on size and shape, and on large white wing patches. Song rather monotonous and scratchy, delivered at a fairly slow pace. **V**

Woodchat Shrike

Woodchat Shrike *Lanius senator* L 19. Breeds in C and S Europe in open, dry country with trees and bushes, sometimes in more wooded terrain. Note dark upperparts with *white shoulder patches and rump*, and striking *rufous-brown crown/nape* (duller in female). Race *badius* (Balearics, Corsica, Sardinia) lacks white wing patch. Juvenile resembles juvenile Red-backed, but has paler upperparts with hint of white shoulder patches and pale rump just visible. Song attractive, full of mimicry but also of scratchy sounds, each phrase often repeated 2–5 times. **V**

Lesser Grey Shrike

Lesser Grey Shrike *Lanius minor* L 20. Breeds in C and SE Europe in open country. Told from rather similar Southern Grey Shrike on smaller size, slightly *shorter tail, longer primary projection*, often more upright posture, *rounder head, lack of white shoulder patches* and in adult summer on *black forehead* (more extensive in male); lacks white supercilium. *White wing patch* prominent in flight. Adult male has *pink breast* in summer. Black forehead to variable degree lost in autumn (in particular in female). Juvenile has whole crown grey barred dark, but underparts are unbarred (unlike in juvenile Great Grey Shrike). Song a peculiar shrill parakeet-like 'tschilip' repeated from exposed song post. Sometimes warbling, scratchy sub-song. Call a Magpie-like harsh, double-note, 'tsche-tsche'. **V**

Great Grey Shrike

Great Grey Shrike *Lanius excubitor* L 24. Breeds in N Europe on open bogs and clearings in taiga, in C Europe on heaths and arid wasteland. Scarce winter visitor to Britain and W Europe in open areas. Easy to spot, perches exposed in top of bush or tree. *Largest* shrike, *gleaming whitish* at a distance. Flight strongly undulating, *white wing patches* conspicuous. Outer part of shoulders also white. Adult has *dark 'mask'* and *white supercilium*. Female finely barred below, male generally not. Often hovers, takes rodents. Pursues small birds in flight. Song is a slow, well-spaced repetition of both harsh and musical notes. Alarm for raptors with an emphatic, grating 'vaaech'. A peculiar ringing call, 'dirrrp' is also given. **WP**

Southern Grey Shrike

266

Southern Grey Shrike *Lanius meridionalis* L 23. In Europe found on arid, rocky heaths in S France and Iberia; wide range in Africa and Asia, where mainly a desert bird. Differs from Great Grey Shrike on *stronger feet, narrower tail* and usually *lack of barred patterns* in juvenile and female. Iberian race (*meridionalis*) has *pink-grey breast, dark grey crown and mantle* and *narrow but distinct white supercilium*. Autumn vagrants in W Europe are often young of Central Asian race (*pallidirostris*), which has *pale bill-base, pale lores*, broadly buff-tipped greater coverts and *very large white wing patch*. Song in Iberia sounds like Great Grey Shrike's. **V**

juv.

♀

♂ **Red-backed Shrike**

Masked

♀

Woodchat

♂ **Masked Shrike**

juv.

♂ **Woodchat Shrike**

Lesser Grey

Great Grey

juv.

♂ **Lesser Grey Shrike**

Southern Grey

♂ **Great Grey Shrike**

♂ **Southern Grey Shrike**

Golden Oriole (family Oriolidae)
A thrush-sized, brightly coloured bird. Nest suspended in fork high up. 3–5 eggs.

Golden Oriole

Golden Oriole *Oriolus oriolus* L 24. Breeds in central and S Europe in groves in cultivated country, prefers deciduous woods with mature trees. Rare breeder in S Britain. Very shy and difficult to get good views of, persistently spends its time in the upper parts of the tree canopy. Appears restless, often on the move. Adult male *bright yellow and black*, female and one-year-old male greenish above, yellowish-white and streaked below. Flight rather like Fieldfare, if anything. Song is a yodelling, mellow as a Blackbird, 'oh-weeloh-**wee**-weeoo' with many variants – the same bird habitually varies the details of the calls. 'Contemplating' males may restrict themselves to quieter 'weeoo' notes. A hoarse and nasal, Jay-like 'kwaa**ek**' is often heard. **SP**

Starlings (family Sturnidae)
Medium-sized, short-tailed and sociable birds. The 4–6 eggs are laid in holes.

Rose-coloured Starling

Rose-coloured Starling *Sturnus roseus* L 21. Breeds some years in SE Europe in open country. Follows the big swarms of locusts, may nest *en masse* in an area for a year or two, then be gone for many years. Rare visitor to NW Europe, mostly in summer. Adult typical, but beware of partially albinistic Starling. Juvenile resembles juvenile Starling, but is *considerably paler, especially on belly, rump and lores*. The bill has yellow base and is not so long and pointed as Starling's. Gregarious, readily associates with Starlings. Resembles Starling in behaviour and calls. **V**

Starling

Starling *Sturnus vulgaris* L 21. Breeds very commonly in cultivated country, especially near human habitations. Makes its home in nestboxes and holes in trees and walls, under roof tiles etc. Note short tail, *speckled plumage* and long, pointed bill. Walks briskly and without pausing over lawns searching for insects (cf. Blackbird). Flight swift, silhouette characteristic with short tail and *pointed wings*. Gregarious. Noisy packs of drab brown juveniles emerge in early summer. Flies in tight flocks, sometimes thousands of birds together in association with roosting in reeds (also city centres) after the breeding season. Call in flight is a short weak buzzing 'tcheerrr'. Alarm a hard 'kyett' (bird of prey) or a grating croak 'stah' (at the nest). The song is varied with whistles, clicking noises and much expert mimicry, recognised by its whining, strained tone and by recurring descending whistles 'seeeooo'. **RW**

Spotless Starling *Sturnus unicolor* L 21. Replaces the Starling in Iberian peninsula, Corsica and Sardinia. Resembles Starling, but *lacks pale spots entirely in summer plumage*. In winter the pale spots are much smaller than the corresponding ones in Starling, and the metallic sheen is slightly weaker. Legs paler pink. Nests colonially. Song resembles Starling's.

Spotless Starling

Waxwing (family Bombycillidae)
A starling-sized bird with a crest on the crown. Sexes alike. Builds open nest in tree. 3–5 eggs.

Waxwing

Waxwing *Bombycilla garrulus* L 18. Breeds sparsely in the coniferous woods of Lapland and eastwards. In winter often seen in flocks in gardens and in city avenues, eats rowan berries. In some years undertakes long-distance movements of invasion proportions. Easy to recognise by the *long crest* and exquisite *cocoa-brown colour* (with tinges of hazel-nut brown and grey). Adult has yellow and white 'V' marks on the primary tips while juvenile has only a whitish straight line. Flight very like Starling's but more regular undulations and slightly slimmer silhouette (no 'shoulders'). Call a clear and high trilling 'sirrr', silvery clear. Song is simple, slow and quiet, consists of the trill call together with harsher notes. **W**

Golden Oriole

♀
(and imm. ♂)

♂

adult

juv.

**Rose-coloured
Starling**

summer

Starling

juv.

winter

**Spotless
Starling**

summer

Waxwing

adult ♂

Spotless Starling   winter

Crows (family Corvidae)

A successful, fairly highly developed group of birds with almost worldwide distribution. The crows and their allies are medium-sized or large, sociable, omnivorous passerines with powerful bills and legs. Raid nests of other birds when chance is offered. Colours most often black, grey and white. Wings rounded. Sexes alike. Have harsh calls for the most part. The 3–7 eggs are usually blue-green and spotted.

Siberian Jay

Siberian Jay *Perisoreus infaustus* L 28. Fairly common to scarce breeding bird in northern coniferous forest (taiga). Sedentary, hardly ever found outside its breeding area. *Fluffy grey-brown with elements of rusty-red*, mainly on the tail. Fearless and inquisitive but also alert. Comes flying noiselessly along and alights right beside camp fire (hoping to get a morsel) or hops around silently inside 'skirts' of spruces, will cling upside down like a gigantic tit. Flies with series of relatively quick wingbeats alternating with glides. The rusty-red colours on rump, tail and wings are clearly visible in flight. Quite silent, but sometimes produces loud outbursts. Rich repertoire of calls. Usual calls are a mewing 'geeah', a harsh 'tchair' and a shrill 'kij, kij'.

Jay

Jay *Garrulus glandarius* L 35. Common in coniferous and mixed woodland, being particularly fond of oak trees. By no means avoids human habitation, but is wary and therefore difficult to see properly. Usually seen on feeding excursions moving from wood to wood and is then recognised by the broad, rounded wings and the laboured flight with rather irregular wingbeats. When glimpsed at closer range in woodland, the *white on rump and wings* is conspicuous (blue wing panel will not be noticed then). Announces itself mainly by calls: most often the typical sudden hoarse shout, 'kshehr', but also the very Buzzard-like mewing, 'peeay'. Also often mimics Goshawk's 'kyek-kyek-kyek-...'. Clucking and intense bubbling sounds given as well. **RWP**

Azure-winged Magpie

Azure-winged Magpie *Cyanopica cyanus* L 35. Strange distribution: Iberian Peninsula and E Asia. Fossil bones found in Spain, so population there is not the result of introduction by man. Found mainly in stone-pine woods. Builds open nest in small colonies. Easy to recognise by *black hood, azure-blue wings* and *long, blue tail*. Back, breast and belly are grey-buff, the *throat white*. Usually seen in small parties. Flight and general behaviour like Siberian Jay's. Active and restless, roaming around a lot, hops about in the crowns of the stone-pines, darts off among the trunks in undulating gliding flight, gleaming azure-blue. Calls include a jay-like, harsh but high-pitched 'zhru*ee*', with faintly rising diphthong, and a clear 'kwee'.

Magpie

Magpie *Pica pica* L 45 (half of which is the tail). Breeds commonly in vicinity of human habitation: in farmland, by settlements, in towns. The domed twig nest is built in the centre of the crown of a deciduous tree, looks like a huge dark ball. Builds a new nest each year. Very alert. The Magpie's reputation as a 'silver thief' is partly undeserved; true, it sometimes takes glittering objects to the nest, but authentic cases of such kleptomania appear to be exceedingly few. The *black and white colour pattern* and *extremely long tail* with green metallic sheen make the Magpie unmistakable. Flight characteristic, with quick, fluttering wing-beats interspersed with short glides. Lives in pairs, but quite large flocks are often seen, e.g. in winter and in association with roosting or good food supply. Apart from the well-known harsh, laughing call, the Magpie makes diverse smacking and plaintive noises and has a quiet song with chirping and twittering sounds. **R**

Siberian Jay

crest raised

Jay

Azure-winged Magpie

Magpie

Nutcracker

Nutcracker *Nucifraga caryocatactes* L 33. Sparse breeder in central European mountains, and in Fenno-Scandia and Russia in lowlands, in coniferous forests with arolla pine or hazel in the vicinity. Arolla pine seeds and hazelnuts are favourite foods, and are cached in autumn. Shy at breeding site. In autumns with poor harvest of cones/nuts, Nutcrackers may invade W Europe, thick-billed European race as well as the much more numerous Siberian slender-billed race *macrorhynchos*. Invasion birds, sometimes in small flocks, are often quite fearless, may be seen in gardens etc. Dark *brown white-spotted plumage* and characteristic flight silhouette with somewhat upslanted body, short tail and rounded wings. Dark tail base contrasts with *white tail tip and undertail area*. Flight desultory and unsteady, reminiscent of Jay's. Call a hard 'rrraah', more rolling than Carrion Crow's, and has a hard, typically hollow and dry ring. **V**

Chough

Chough *Pyrrhocorax pyrrhocorax* L 40. Breeds in mountain districts and along steep rocky coasts in S and W Europe, respectively. *Glossy black* with *long and curved red bill*. Juvenile's bill brownish-yellow. The wings have short 'arm' and long 'hand', are broad and have *6 clearly spread, flexible 'fingers'* (much more obvious than Alpine Chough's). Tail rather short, square. Black underwing-coverts, darker than flight feathers. Often plays in groups in the air in front of rock faces. Superb aerial acrobat, dives headlong at breakneck speed with closed wings. Call a characteristic 'keeach' (in all likelihood origin of English name), basically like Jackdaw but considerably hoarser, and 'thicker' and more 'caustic'. **R**

Alpine Chough

Alpine Chough *Pyrrhocorax graculus* L 38. Breeds in S Europe in high mountains up to the snow line. Flocks are often seen circling around the summits and searching for food on alpine meadows high above Jackdaw ground. Can in winter be seen in villages in the valleys, but are even then usually met with at the top stations of ski resorts. Nests in colonies in rock crevices and caves. Much resembles Chough, but has shorter, *yellow bill*, not such glossy plumage, decidedly *less prominent 'fingers' at the wingtips, longer tail*. Pink legs (juvenile has darker legs). Note that juvenile Chough has yellowish bill, too (brown-toned). Distinguished from Jackdaw even at a distance by more obviously fingered wingtips, slightly longer tail with narrower base, and also in the right lighting conditions by black underwing-coverts, distinctly darker than grey flight feathers (Jackdaw an even-grey below). Commonest calls are very peculiar and characteristic: a clear, piercing 'tzii-eh' (sharp beginning, whining). Also a rolling 'krrrrül' (something like a noisy young Starling in nest hole). Both calls have a particular 'electric' quality, reminiscent of sound produced by throwing a stone on to thin fresh ice. Other minor calls are more akin to Chough calls.

Jackdaw

Jackdaw *Corvus monedula* L 33. Breeds commonly in cultivated country, in older deciduous wood, in towns and in rocky mountains and coastal cliffs. Nests in hollow trees, chimney stacks etc, cliffs, with tendency towards colonial breeding. Black with *grey nape. Iris greyish-white*. Flight powerful and fast, always quicker and with deeper wingbeats than Carrion Crow's, may somewhat recall that of the pigeons. Is very sociable, almost always seen in pairs or in flocks; division into pairs obvious even in the flocks, which otherwise are much more closely packed together than, e.g. those of Carrion Crow. Feed on fields, often together with other crow species and Starlings. Are often seen circling in flocks at quite high altitude, playing tag and performing acrobatic tricks. Gather in autumn to roost in certain selected towns; remarkable mass flight display as dusk approaches, and loud, noisy cackling at twilight. The Jackdaw's usual calls are a loud, jarring, nasal 'kye' and a more drawn-out 'kyaar'. When giving alarm against birds of prey, a hoarse 'cheehr'. **RSW**

Nutcracker

juv.

adult

Chough

Alpine Chough

Jackdaw

Rook

Rook *Corvus frugilegus* L 46. Typical bird of cultivated lowlands, where it nests in colonies (rookeries) in groups of trees and small woods. Common over whole of Britain and Ireland except N Scotland. Nests are placed close together high in the tops of trees. Gregarious all year. Feeds in flocks in fields, performs aerial games in flocks above the nest trees. *Black plumage with violet sheen*, comparatively long and evenly tapering bill and, in adult, *a pale bare area where bill joins feathering* are characteristic. Peaked crown and bushy 'trousers' are normally apparent, but not invariably. Wings slightly longer and narrower than Carrion Crow's, flight more elegant with slightly deeper, more elastic wingbeats; also glides more. Immature is duller and does not have bare skin at bill-join (may retain feathering there until one year of age), is therefore extremely like Carrion Crow. Distinguished, however, by *bill shape*, by *pale mouth flanges* and also by call (differences in flight certainly exist but difficult to judge in case of single individual; odd birds may give wrong impression). Call more nasal and hoarser than Carrion Crow's, not so open, coarse and rolling, 'kah'. **RW**

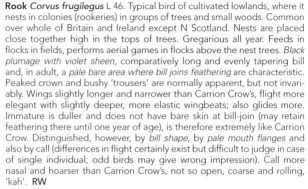

Carrion Crow

Carrion Crow *Corvus corone* L 46, W 85. Common in W Europe in all types of fairly open country, but also in towns and cities. Builds open twig nest in treetop in clump of trees, well concealed, not exposed as Magpie. Never breeds in colonies, is a territorial bird. Black with *moderate, blue sheen*. Best told from Rook by black feathering at bill base (though juvenile Rooks also have this), *stouter bill more downcurved at tip* and by much coarser, less nasal croaking call. *Mouth flanges always dark*. Does not have 'trousers'. *Tail slightly shorter and not so rounded at the tip*. Outer wing on average slightly shorter than Rook (Rook can look like Raven in silhouette, which Carrion Crow hardly does). Also has slightly less elastic, more listless wingbeats than Rook. Distinguished from Raven by smaller size, smaller bill and square-cut tail. Gathers in flocks outside breeding season, in search of food and for roosting. Omnivorous, nest-raider in summer, often eats refuse and carrion. Call a harsh croaking 'kraa' often repeated several times. In internal quarrels and also when warning of 'harmless' birds of prey give persistent, grating 'krrrr' calls. (Superior birds of prey are showered with furious 'kraa' calls.) **RW**

Hooded Crow

Hooded Crow *Corvus cornix* L 46. The grey and black crow of NW Scotland, Ireland, the Isle of Man and the eastern half of Europe is closely related to the Carrion Crow, but is now generally separated as a different species. Breeds in cultivated country, commonly but in single pairs. Wary, well-used to man. Easily recognised by grey and black plumage. In habits resembles Carrion Crow, with which it often associates and also interbreeds where the two meet. Flight as Carrion Crow. Flies singly or in loose formation. Hooded Crows from NE Europe visit east coast of Britain in winter, mainly Oct–Apr. Voice like Carrion Crow's. **R**

Raven

Raven *Corvus corax* L 65, W 125. Breeds fairly commonly on rocky coasts and mountains and in extensive woodland. Maintains lifelong pair bonds. Also outside the breeding season is very often seen in pairs; two dots moving along a ridge are often Ravens. But may sometimes gather in quite large parties. The nest is placed on cliff shelf or in tree. An early breeder, often incubating in Feb–Mar. Feeds on small animals, carrion and refuse. Roams widely, visits refuse tips, slaughterhouses and similar. *Largest* of the passerines, clearly bigger than Buzzard. *All-black* plumage, powerful bill and *long, wedge-shaped tail* distinguish it from the smaller crow species. Flight with quite measured but very driving wingbeats. Soars more often than other crows. May then be confused with birds of prey, but Raven never holds its wings raised in soaring flight. Often performs half-rolls, also in regular cruising flight. Shy and wary. Call 'prruk', deep, resonant; alarm 'rak-rak-rak'. In spring various clucking noises. **R**

Rook

juv.

Carrion Crow

Hooded Crow

Raven

Sparrows (family Passeridae)

Sparrows are small, thick-billed birds, more dumpy than most other passerines. In some species the sexes are alike. Sociable, most breed colonially. Feed on the ground. Lack well-developed singing ability. Nest in holes or build domed twig nests. Clutches of 4–8 eggs. The Snow Finch belongs to this family, but is treated on p. 296 for easier comparison with Snow Bunting.

House Sparrow

House Sparrow *Passer domesticus*. L 14.5. Closely associated with human habitation, breeds commonly around farmyards, in towns and villages. Pronounced resident. Has declined markedly in recent times. Sociable. Nests under roof tiles, in holes, etc. (may also build nest in the open: large, domed). Gathers in hedges and bushes in loud noisy flocks. Small flocks flies as one body, straight and seemingly with some effort. Rather heavy and ungainly. Male on closer inspection quite dainty: *crown and rump grey, 'temples' chestnut-brown, large black 'bib'*. Female and juvenile are more uniformly drab grey-brown, though note pale buff-brown streak behind eye. Calls are simple and rolling or slightly impure, 'trilp', 'chev', 'chierp', etc. Alarm is a crude rattling 'cherrrr-r-r-r', slowing at the end. Song consists of 'trilp' calls repeated boldly at different pitches. **RWP**

Italian Sparrow *Passer (domesticus?) italiae* L 14.5. A controversial form of sparrow living in S Switzerland, N and C Italy, Corsica and Crete, by some regarded as a race of House Sparrow, by others as a race of Spanish Sparrow, or as a stable hybrid population between the two. The male shares with Spanish Sparrow the *entirely chestnut-crown*, but underparts are more like in House Sparrow, but with *cleaner white cheeks*. Behaviour and calls are the same as in House Sparrow.

Spanish Sparrow

Spanish Sparrow *Passer hispaniolensis* L 14.5. Despite its name this species has a rather restricted range in Spain but is abundant in N Africa and locally common in Sardinia, SE Europe and Turkey. Breeds colonially in trees in open, cultivated country without any association with villages, often in windbreak trees, riverine poplars, etc. May also breed in stork nests. Is partly a migrant, at least in the northern part of the range; large flocks can be seen moving during migration seasons. Male is easily distinguished from male House Sparrow by *larger black 'bib', black streaking on flanks, much black on back* and *whitish cheeks* together with entirely chestnut-brown crown. Has short *white supercilium*. Extent of black varies, most are generally much blacker than Italian Sparrows, others are not that different. Female and juvenile are often indistinguishable from House Sparrow, but have on average a trifle more well-marked greyish streaks on flanks, slightly heavier bill and whiter belly. Call and song resemble House Sparrow's, but are slightly louder and deeper. **V**

Tree Sparrow

Tree Sparrow *Passer montanus* L 14. Breeds commonly in parks, gardens and cultivated country with trees. Less tied to human habitations than House Sparrow and can be found breeding in pure woodland. Nests in holes in trees or houses, also in nestboxes. Smaller and slimmer than House Sparrow. Note *vinaceous-brown crown, small 'bib'* and *a small black spot on whitish cheek*. Thin whitish semi-collar. Sexes alike. Some calls resemble House Sparrow's, others are more distinctive, like a merry, slightly nasal, disyllabic 'tsoo**wit**', and nasal but dry 'tett' which can be run into short series in flight, 'tett-ett-ett-ett'. Song a series of chirps. **RWP**

Rock Sparrow

Rock Sparrow *Petronia petronia* L 16. Breeds locally in S Europe in rocky country, cultivated as well as wild. Sometimes nests in ruins, within towns or in hollow trees in orchards. Resembles female House Sparrow with *broad pale supercilium*, but has a *pale crown-stripe* and *whitish tail band* (noticeable mostly in flight). *Lower mandible pale with dark tip*. Small yellow spot on breast not very conspicuous, shows mostly when bird puffs up its feathers. Much more active than House Sparrow. Usually seen in small loose flocks. A variety of short calls like 'wed' and '**dlee**yu'. Song contains a characteristic very sweet 'pee**uh**-ee'. **V**

House Sparrow

♀ ♂

Italian Sparrow

♂ ♂

Spanish Sparrow

♀

Tree Sparrow

juv.

Rock Sparrow

Finches (family Fringillidae)

Finches are small with short, heavy bills. Seed-eaters. Males in particular are often brightly coloured. Nest in trees, shrubs, etc. Build cup-shaped nests. Clutches of 3–6 eggs.

Chaffinch

Chaffinch *Fringilla coelebs* L 15. One of Europe's commonest birds. Nests in woods, in deciduous as well as coniferous, in gardens and fields. The nest is most often in a tree fork, skilfully constructed and well camouflaged with moss and lichen. Migrant in north and northeast, large movements in early October. In winter common in flocks in fields, usually rather close to woods. Much drawn to beech woods. Male brightly coloured, female and juvenile dull grey-brown with greenish tinge; they all have prominent *white wingbars*, white on outer tail-feathers and green rump. One of the most zealous singers in the wood, song loud. From well-visible song post gives a tuneful, short but forceful rattle which drops down the scale and ends with a flourish, e.g. 'zit-zit-zit-zit-sett-sett-sett-chitter**eee**-dia'; sometimes a woodpecker-like 'kik' is added to the final flourish. Also often repeats a loud call persistently, a call that varies geographically: 'hu**itt**' in N Europe, 'rrhü' in C Europe (often referred to as the 'rain-call') and 'heep' (like alarm of Thrush Nightingale) in S Europe. Alarm is an eagerly repeated 'chink'. Flight call a short, low 'jup, jup'. **RWP**

Brambling

Brambling *Fringilla montifringilla* L 15. Breeds commonly in taiga and subalpine birch forest. On migration and in winter frequents woods, parks, gardens and fields, usually in flocks. Passage flocks tighter than migrating Chaffinch flocks. Note *white rump, dark back* and *white wingbars*. Flocks are often mixed with Chaffinches. In some 'invasion years' flocks of thousands and even millions can occur in areas of beech woods; eats beechnuts. Song is a dreary, monotonously repeated, rolling and wheezing 'rrrrhee' (at a distance slightly reminiscent of a cross-cut saw). Sings from well-exposed song post. Flight call 'jek jek', like Chaffinch but more nasal and a little harder. Call is a characteristic nasal 'teh-ehp'. Alarm call at breeding site a hard ringing 'slitt, slitt'. **WP**

Bullfinch

Bullfinch *Pyrrhula pyrrhula* L 16. Breeds fairly commonly in deciduous as well as coniferous wood (taiga), also in gardens and orchards. *Underparts red in male*, greyish-brown in female. Both have *black cap*, grey back, *white rump* and *white bar on black wings*. Bill short and heavy. Juvenile is pale brown with no black on face other than peppercorn eye. Quiet. In winter sometimes in small flocks, incl. in gardens, in summer very shy and withdrawn. Eats berries, buds and seeds. Call a soft, sad, melodic 'pew'; when perched a more downwards inflected 'pee-u'. Some northern birds (from Finland and Russia?) have nasal, buzzing call, like toy trumpet instead of 'pew'. Song weak, unmusical: calls mixed with wheezing notes. Often includes two calls in rapid succession, 'pew-pew'. **RWP**

Bullfinch *juv.*

Hawfinch *Coccothraustes coccothraustes* L 18. Breeds sparsely in deciduous and mixed wood. Attracted to cherry orchards, easily splits kernels with its colossal bill. Hornbeam seeds also form a popular food. Easy to overlook as it is shy, rather silent and spends most of time in foliage of treetops. Note *very large and powerful bill* (blue-grey in breeding season, yellowish-white rest of year), short tail and *broad white wingbar*, characters which are well-visible in flight. Female is basically like male but has pale grey secondary panel on folded wing, is less warm and rich in colours. Apart from the white bars on wing and tail, juvenile is a rather nondescript yellowish-brown bird (lacks among other things black markings on face). Can be sexed by the secondaries character. Call a metallic, very hard 'pix!' like a powerful, single Robin 'tic', a Blackbird-like 'srree', and 'chi' with harsh quality of Spotted Flycatcher. Song consists of low, strained 'tee-eeh' notes mingled with the calls. **R**

Hawfinch

♂ winter

♀

♂ summer

Chaffinch

♂

♀

♂ summer

Brambling

winter

♂

♂

♀

Bullfinch

♂

adult

Hawfinch

juv.

♂

279

Citril Finch

Citril Finch *Serinus citrinella.* L 12. Breeds in C and SW Europe in mountains with coniferous woods. Its presence depends on spruce seeds, but it also feeds on dandelion seeds from alpine meadows. In winter descends to lower level. *Unstreaked, greenish-yellow underparts and rump, unstreaked greyish-green back, yellowish wingbars* and ash-grey nape distinguish adult. Juvenile is browner, with faint yellow tinge on belly, and heavily streaked. Usually seen in small parties. Call a fast, 'trembling', nasal 'keke-ka', or single 'ke', with a nice cracked ring to it. Has also more Siskin-like calls. Alarm an emphatic, hoarse 'teehah'. Song fast, twittering and a bit 'disorderly', recalling the song of Goldfinch, often given in song flight. **V**

Corsican Finch

Corsican Finch *Serinus (citrinella) corsicanus* (Not illustrated.) L 12. Until recently regarded as a race of Citril Finch, now generally separated as a species in its own right. Confined to Corsica and Sardinia, where it is a resident, often found among junipers, shrubs and heather both in mountains and lower down. Very similar to Citril Finch but adult differs in having *tawny-brown back with some dark, diffuse streaking.* Male has somewhat more yellow face and underparts, not so green-tinged as in Citril Finch. Calls same as in Citril, but song more structured, a little like Wren's.

Serin

Serin *Serinus serinus* L 11. Common over most of the Continent, in Britain rare and very local in summer (irregular breeder). Found in parks, coniferous belts, gardens, often in villages. Streaked plumage, largely *yellow underparts and rump.* In summer, when greyish-green feather fringes are worn away, head can look almost all-yellow. Juvenile lacks yellow and is streaked; most easily recognised by *small size* and *very short bill.* Often feeds on the ground. Common call a high-pitched, metallic twittering, 'zr-r-litt'. Sings frequently from high and exposed song post or in Greenfinch-like song flight with slow wingbeats, a jingling, chirping and harsh twitter at fast tempo (somewhat reminiscent of glass being ground). **SP**

Greenfinch

Greenfinch *Carduelis chloris* L 14.5. Breeds quite commonly in open, cultivated country with dry, bushy patches; also found in gardens and parks. Frequent visitor to birdtables in winter. Adult male is an *attractive yellowish-green* below and olive-green above (colours brightest in summer). Female has slightly duller and more greyish-green colours, while juvenile is greyish-green and brown and *streaked.* All plumages are characterised by *bright yellow on outer tail-feathers and on edges of primaries.* Heavy in build, has *quite large head and heavy bill.* Fast flight with longer and deeper undulations than, e.g. Chaffinch. Call in flight a rapid rolling 'djururut' or just short 'djup' notes, more emphatic than Chaffinch. When perched often gives a slightly hoarse Canary-like alarm, 'dyuwee'. From elevated perch (treetop, TV aerial) gives a powerful, trilling song which may also be delivered in butterfly-like song flight. One song variant, which may be woven into the trilling song as well as being delivered on its own at short intervals, is a loud drawn-out wheezing 'djeeeesh' (vaguely recalling song of Brambling). **RSWP**

Siskin

Siskin *Carduelis spinus* L 12. Breeds fairly commonly but locally in coniferous forest, especially tall spruce. Numbers vary with seed production. Smallest finch of N Europe. *Greenish and streaked,* has yellow tail-sides and wingbars. Male has *black forehead and chin patch.* Outside breeding season gathers in dense flocks (often mixed with Redpolls), feeds silently, hanging upside-down in birches and alders, suddenly launches out in dense swarm in short circular trip uttering buzzing twitter. In Britain often visits garden nut feeders in winter. Flight light, in long, deep undulations. Call a drawn-out, clear, sad 'dlu-ee' or 'dlee-u', as well as a dry 'kettekett'. Song a fast, twittering chatter, ending with a feeble wheeze. Performs song flight. **RWP**

juv.

Citril Finch

adult

♀

♂

Serin

♀

♂

Greenfinch

Greenfinch display flight

♂

♀

♂

Siskin

281

Goldfinch

Goldfinch *Carduelis carduelis* L 14. Breeds fairly commonly in open country at woodland edge, also often seen in parks and gardens. Easily overlooked during breeding season. Adult easily recognised by *red face* with rest of head black and white. Wing black with a *broad bright yellow bar*, particularly striking in flight. *Rump white*. Sexes are impossible to separate in the field. Juvenile has wing and tail feathers as adult's (broad bright yellow wingbars on outerwing), but is otherwise insipid grey-brown with dark streaking. Last to be moulted is the head, and so in early autumn immature Goldfinches are seen which look like adults except for the pale grey head. Sociable outside breeding season and often occurs in small flocks. Specialised on seeds of thistles; in autumn and winter is usually found on open 'untidy' meadows with many thistles. Call a characteristically sharp and high-pitched 'ticke**lit**'; rasping Sand Martin-like calls are also heard from larger flocks. Song resembles both Greenfinch's and Siskin's, recognised by the call being mixed in, as well as attractive mewing sounds. **RS**

Linnet

Linnet *Carduelis cannabina* L 13. Breeds commonly on heaths and in open country with hedges and bushes, in parks and gardens. Very often found in pairs. In late summer gathers in large flocks on ripe rape fields, stubble etc., often together with Greenfinches. The Linnet is a fairly small bird. Male easily identified by *grey head, raspberry-red on breast and forehead, hazelnut-brown back* and white on edges of primaries and tailfeathers. Bill grey. In autumn the red is more softened by dull fringes. Female and juvenile lack the red, have streaked breast, less pure colours and can therefore be confused with Twite. Risk of this, however, occurs mainly in autumn and winter, and then the Linnet's dark bill immediately gives clear indication. Also, the *white on edges of primaries are much more striking*, while on the other hand wingbars are virtually lacking, back and breast have weaker streaking, and *throat is greyish-white* (with fine streaking in centre) instead of (unstreaked) buffish-yellow. In addition, both sexes of Linnet has an altogether particular facial expression owing to the light area on centre of cheeks, surrounded by dark. Is voluble and has many calls. In flight a rather nasal ricocheting 'kne**tett**', or just 'tett', sometimes combined with short trills or thin soft whistles, e.g. 'peeuu', 'trrrü' and 'tukeeyü'. Gives very varied and pleasing twittering song from well-visible song post. **RSW**

Twite

Twite *Carduelis flavirostris* L 13. Breeds on upland moors, also near coast. In winter on cultivated lowlands, rubbish tips, along flat coasts and similar places, where seeks food in flocks among tall weeds. Fearless but restless and therefore difficult to see properly, often flies a long way off. In general most resembles Redpoll and Linnet. Often associates with both. Compared with Redpoll, longer-tailed, more *yellowish-brown in tone* and *lacks dark areas on forehead and throat*; resembles female Linnet, but *bill is bright yellow* (though in summer grey-brown, as in female House Sparrow), cheeks and breast yellower brown, *throat unmarked buffish* (not whitish and streaked), the wing has a Redpoll-like pale bar (tips of greater coverts), does not have such obvious white on outer primaries as Linnet. Legs black. Male has pink on rump, though this is very difficult to discern in the field, especially in autumn/winter (when brown fringes conceal). Most typical call is a nasal, hoarse, fine '**twe**it' (hence the name), which is totally characteristic and is heard in particular from large flocks. Also has a more conventional chattering call 'jek, jek', which is something in between calls of Common Redpoll, Brambling, Linnet and Greenfinch but still quite distinctive. Song chattering, with calls intermingled. **RSW**

Goldfinch

juv.

adult

Linnet

♀

♂

♂

Twite

♀

♂ summer

♂

Redpoll

Lesser Redpoll

Arctic Redpoll

Trumpeter Finch

Common Rosefinch

Pine Grosbeak

Common Redpoll (Mealy Redpoll) *Carduelis flammea* L 12.5. Breeds in Fenno-Scandia and Russia in subalpine birch zone, but also in coniferous forests. Depending on supply of birch seeds may make large movements west and south in the winter. A small finch with *mainly grey* colours (tinged warmer buff-brown when fresh in autumn). Told from all other species (save the other redpolls) by *red fore-crown* (attained from Aug onwards by first-years) and *small black chin patch*. Adult male *pale red on breast and rump* (red becoming darker in summer), rump sometimes almost unstreaked. A large, darker brown and heavy-streaked race breeds in Greenland (*rostrata*), of which a few reach Britain annually. Sociable, often nests in loose colonies. In winter in dense flocks in birches and alders, though also feeds in tall weeds like Twite. Call 'chut, chut-chut, chut…', cracked, with metallic echo. A hoarse Greenfinch-like 'dyuee' is often popped in, especially in anxiety. Flight song consists of the call, alternating with a drawn-out, dry, trilling rattle: 'chut chut chut serrrrrrr chut chut chut…'. **WP**

Lesser Redpoll *Carduelis (flammea) cabaret* L 11.5. A smaller form of Common Redpoll (regarded as a separate species in Britain, but as a race in many European countries). Breeds in Britain and in C Europe east to the Alps; has in recent decades expanded into S Scandinavia. Found in woods and stand of trees in open terrain such as seashore meadows. *Smaller, darker* and *more tawny-brown* in fresh plumage than Common Redpoll, and *streaking of upperparts and flanks usually heavier*. Typical birds can be picked out in the field, but many are inseparable from Common. Voice very similar to Common Redpoll's. **RWP**

Arctic Redpoll *Carduelis hornemanni* L 12.5. Breeds sparsely in the subarctic region, in the willow zone (rarely upper birch forest). In winter seen rarely among Redpolls. Very like Redpoll. Distinguishing marks are *unstreaked, white rump* (particularly characteristic of adult male; first-years at least may have faint streaking on rump), fewer and narrower streaks on flanks and *paler head* (in autumn tinged yellowish-buff) *and back*. On average shorter bill and looser plumage than Redpoll, but differences subtle. In practice many 'pale redpolls' must be left unidentified (even trapped ones can be difficult). Calls similar to Lesser Redpoll's. **V**

Trumpeter Finch *Bucanetes githagineus* L 14. N African and W Asiatic species, also in SE Spain (Almeria). *Bill characteristic: very short and stubby*. Male in summer brownish-grey with *rose-pink tinges, pale red bill*. In winter more like female which is somewhat duller in colour. Song a peculiarly nasal, monotonous and very loud buzzing ('toy trumpet'). **V**

Common Rosefinch *Carpodacus erythrinus* L 14. Colonised N Europe from the 1950s. Inhabits luxuriant scrub with tall herbs and scattered trees. Arrives very late in spring. Male in full summer plumage *strawberry-red on head, breast and rump*. Female and juvenile insipid grey-brown with hint of streaking and faint wingbars (cf. females of House Sparrow and Linnet). *Stout, heavy bill*. One-year-old males breed in female plumage. Attracts attention mostly by its song, a short, sharply whistled phrase, 'weeje-wü weeja' or 'pleased to meet you' etc. (the theme varies), typically pure and soft in voice. Call a fresh, pure 'ueet', with same voice as in song. Alarm an almost Greenfinch-like 'jay-ee'. **P**

Pine Grosbeak *Pinicola enucleator* L 20. Breeds sparsely in northern taiga, may go up into subalpine birch forest. Feeds on buds up in the spruces, searches for berries when hopping on ground. Quiet, easy to overlook. Male *raspberry-red*, female and immature male mustard yellow and dust-grey, all with *two distinct white wingbars*. In some winters moves south and west in large numbers, plundering the rowans, amazingly fearless. Big and sturdy, obviously long-tailed. In flight resembles small Fieldfare, but flight more undulating. Call a clear, strongly fluting 'peelee-jeeh, peeleejü'. Conversational call a subdued 'büt, büt'. Song a fast, crystal-clear series, rather like Wood Sandpiper display. **V**

Common Redpoll ♂

Lesser
Redpoll ♂

Arctic Redpoll ♂

♀

♂ winter

Trumpeter Finch

♀

♂ summer

Common Rosefinch

♀
(and imm. ♂)

♂

Pine Grosbeak

♂

♀

285

Crossbills (family Fringillidae, genus *Loxia*)

Live on conifer seeds. Specialists in cutting off and prising open cones (each species its preferred sort) with their powerful, crossed bills. Breeding usually takes place in late winter (during severe cold), since the seeds of the cones (the food of the young) ripen at that time. Breeding may, however, occur also in summer or early autumn. Crossbill populations lead a roving existence, descending on those regions which have a rich cone production just in that actual year. In years of cone failure mass long-distance movements take place.

Scottish Crossbill *Loxia scotica* L 16.5. Sedentary breeding bird in Scotland, in pine forest, and Britain's only endemic bird. However, its taxonomic status as full species has been questioned. Between Parrot Crossbill and Common Crossbill, has intermediate dimensions and calls. Plumages the same as for the other two species. Often impossible to distinguish from its relatives in normal encounters. **R**

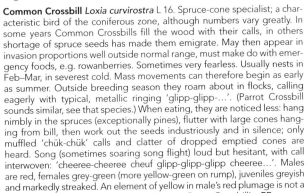

Common Crossbill

Common Crossbill *Loxia curvirostra* L 16. Spruce-cone specialist; a characteristic bird of the coniferous zone, although numbers vary greatly. In some years Common Crossbills fill the wood with their calls, in others shortage of spruce seeds has made them emigrate. May then appear in invasion proportions well outside normal range, must make do with emergency foods, e.g. rowanberries. Sometimes very fearless. Usually nests in Feb–Mar, in severest cold. Mass movements can therefore begin as early as summer. Outside breeding season they roam about in flocks, calling eagerly with typical, metallic ringing 'glipp-glipp-...'. (Parrot Crossbill sounds similar, see that species.) When eating, they are noticed less: hang nimbly in the spruces (exceptionally pines), flutter with large cones hanging from bill, then work out the seeds industriously and in silence; only muffled 'chük-chük' calls and clatter of dropped emptied cones are heard. Song (sometimes soaring song flight) loud but hesitant, with call interwoven: 'cheeree-cheeree cheuf glipp-glipp-glipp cheeree...'. Males are red, females grey-green (more yellow-green on rump), juveniles greyish and markedly streaked. An element of yellow in male's red plumage is not a certain sign of immaturity. *Bill not so heavy* as Parrot Crossbill's. **RP**

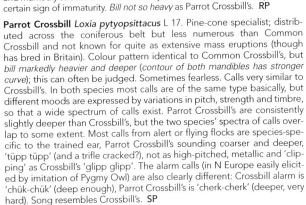

Parrot Crossbill

Parrot Crossbill *Loxia pytyopsittacus* L 17. Pine-cone specialist; distributed across the coniferous belt but less numerous than Common Crossbill and not known for quite as extensive mass eruptions (though has bred in Britain). Colour pattern identical to Common Crossbill's, but *bill markedly heavier and deeper* (*contour of both mandibles has stronger curve*); this can often be judged. Sometimes fearless. Calls very similar to Crossbill's. In both species most calls are of the same type basically, but different moods are expressed by variations in pitch, strength and timbre, so that a wide spectrum of calls exist. Parrot Crossbill's are consistently slightly deeper than Crossbill's, but the two species' spectra of calls overlap to some extent. Most calls from alert or flying flocks are species-specific to the trained ear, Parrot Crossbill's sounding coarser and deeper, 'tüpp tüpp' (and a trifle cracked?), not as high-pitched, metallic and 'clipping' as Crossbill's 'glipp glipp'. The alarm calls (in N Europe easily elicited by imitation of Pygmy Owl) are also clearly different: Crossbill alarm is 'chük-chük' (deep enough), Parrot Crossbill's is 'cherk-cherk' (deeper, very hard). Song resembles Crossbill's. **SP**

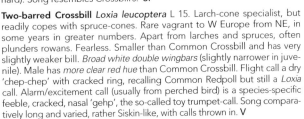

Two-barred Crossbill

Two-barred Crossbill *Loxia leucoptera* L 15. Larch-cone specialist, but readily copes with spruce-cones. Rare vagrant to W Europe from NE, in some years in greater numbers. Apart from larches and spruces, often plunders rowans. Fearless. Smaller than Common Crossbill and has very slightly weaker bill. *Broad white double wingbars* (slightly narrower in juvenile). Male has *more clear red hue* than Common Crossbill. Flight call a dry 'chep-chep' with cracked ring, recalling Common Redpoll but still a *Loxia* call. Alarm/excitement call (usually from perched bird) is a species-specific feeble, cracked, nasal 'gehp', the so-called toy trumpet-call. Song comparatively long and varied, rather Siskin-like, with calls thrown in. **V**

Scottish Crossbill
♂

Common Crossbill
♂

juv.

♀

♀

juv.

Parrot Crossbill
♂

Two-barred Crossbill

♀

♂

juv.

Buntings (family Emberizidae)

Buntings are rather small birds, with short, thick bills with slightly S-shaped cutting edges. Live in open country with bushes, often in cultivated districts with hedges and windbreak trees but also in upland tracts on bare moorland and in willow or in reeds and damp thickets. Only Rustic Bunting and perhaps Little Bunting thrive in more enclosed woodland. In spring and summer often noticed by species-specific song. Males have brighter plumages than females. Juveniles resemble females but are duller and more streaked (see pp.298–299). In some species sexes are alike. Outside breeding season often seen in flocks. Feed mainly on seeds on ground, but nestlings are often fed on insect larvae and the like. Nest on ground or in low bushes. The 3–6 eggs are intricately patterned.

Corn Bunting

Corn Bunting *Emberiza calandra* L 18. Breeds fairly commonly in C and S Europe in open, cultivated country, in S Europe often also on dry mountain slopes without taller vegetation. Marked decline in northern parts of range. Most males are polygamous. Outside breeding season seen in flocks. *Grey-toned brown*, heavily streaked without particular identification features. Streaks on throat often merge to *dark patch on sides* or centre. Lacks white on tail-sides. Legs pinkish. Sexes alike. Looks *bulky*. Flight heavy. Even at a distance easily distinguished from Skylarks in flight, also by *lack of pale trailing edge to wing*. Often flies with *legs dangling*. Call a hard, sharp, almost clicking 'tik'. Also a loud, hard grating 'strri(e)', persistently repeated. Usually sings from well-visible perch, often from telegraph wire. Song metallic, a monotonously repeated accelerating jarring sound, 'tük tük zik-zee-zrrississ'. **RW**

Little Bunting

Little Bunting *Emberiza pusilla* L 13. Breeds in far NE Europe, including in NE Finland but irregular farther west. Prefers damp, fairly open wood. Very rare migrant southwards. In summer plumage recognised by *reddish-brown head* (adult male also chin) with black stripes on sides of crown (*pale, reddish-brown stripe down centre of crown* characteristic) together with narrow black lines partially framing rusty-brown cheek. Despite smaller size can be confused with juvenile Reed Bunting in autumn, but recognised by *evenly rusty-yellow-brown cheek patch* (not dark and variegated black-brown-white) on which the bordering black streak below does not reach bill, proportionately *slightly longer bill* with *straight* (not convex) culmen, *dull brown* (not reddish-brown) *lesser wing-coverts above* (not always seen on perched bird), paler feet and usually narrower, shorter and blacker streaks on underparts. Usually distinct *pale eye-ring* and *pale buffish-white covert bar*. Call a hard, sharply clicking 'zik', like Hawfinch call in miniature. Song fairly quiet, varying, contains several motifs recalling Ortolan and Reed Buntings among others. **V**

Rustic Bunting

Rustic Bunting *Emberiza rustica* L 14.5. Has until recently been fairly common in Swedish and Finnish taiga, but marked decline recently. Inhabits taiga, also birch forest, bounded by small bogs and streams. Migrates mainly due east to reach SE Asia, but annual vagrant (usually in autumn) to Britain. Male in summer plumage has characteristic black and white markings on head. Also characteristic are *reddish-brown breast band, reddish-brown* flanks together with white belly. Female and autumn birds most easily confused with Reed Bunting, but latter is dark-streaked on breast and flanks, lacks red-brown colour; Rustic Bunting's *rump is reddish-brown*, Reed Bunting's brownish-grey; bill has straight culmen in Rustic, convex in Reed; more prominent light wingbars in Rustic, and legs paler (pink). Call very like Song Thrush's but slightly higher and more distinct, 'zit'. Song clear and melodic, fairly short, has mournful ring like Lapland Bunting and is irresolute like Dunnock's in character, e.g. 'dudeleu deluu-delee', very mellow in tone. **V**

Corn Bunting

Little Bunting

Rustic Bunting

♀ *summer*

♂ *summer*

1st-winter

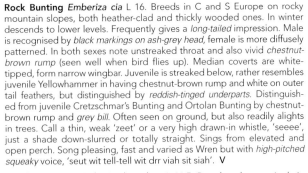

Rock Bunting

Rock Bunting *Emberiza cia* L 16. Breeds in C and S Europe on rocky mountain slopes, both heather-clad and thickly wooded ones. In winter descends to lower levels. Frequently gives a *long-tailed* impression. Male is recognised by *black markings on ash-grey head*, female is more diffusely patterned. In both sexes note unstreaked throat and also vivid *chestnut-brown rump* (seen well when bird flies up). Median coverts are white-tipped, form narrow wingbar. Juvenile is streaked below, rather resembles juvenile Yellowhammer in having chestnut-brown rump and white on outer tail feathers, but distinguished by *reddish-tinged underparts*. Distinguished from juvenile Cretzschmar's Bunting and Ortolan Bunting by chestnut-brown rump and *grey bill*. Often seen on ground, but also readily alights in trees. Call a thin, weak 'zeet' or a very high drawn-in whistle, 'seeee', just a shade down-slurred or totally straight. Sings from elevated and open perch. Song pleasing, fast and varied as Wren but with *high-pitched squeaky* voice, 'seut wit tell-tell wit drr viah sit siah'. **V**

Ortolan Bunting

Ortolan Bunting *Emberiza hortulana* L 16.5. Breeds rather sparingly in open, cultivated country (lowland) with clumps of trees (N Europe) or in open mountainous terrain with scattered trees (S Europe). Marked decline in NW Europe. Adult male distinguished from other buntings by *greenish-grey, unmarked head, pale yellow throat* and greenish-grey breast band. At close range *narrow yellowish-white eye-ring can be seen*. Legs and *bill* pinkish. Outer tail-feathers edged white. Distinguished from Cretzschmar's Bunting by yellow (not orange) throat and greenish-grey (not blue-grey) head and breast band. Female similar to male but has brown-toned greenish-grey head with dark streaking and distinctly streaked greyish breast band. Juvenile is brown and streaked as a pipit, has grey-toned brown, streaked rump (not reddish-brown as in Yellowhammer and Rock Bunting). Call a clear, metallic 'slee-e'; also heard from night migrants in Aug, repeated loosely (3–5 times per flight over). A contact call, heard, e.g. from diurnal migrants, is a muffled, dry 'plett'. When agitated a short 'chu' (or '**slee**-e' and 'chu' alternating at 2-sec. intervals). Song varies individually, but ringing tone is specific, and typically the second part has lower notes than first: 'swee swee swee swee drü drü' or 'drü drü drü seea seea'. In S Europe the second part usually is only one cracked, falling note (cf. Cretzschmar's). **P**

Cretzschmar's Bunting *Emberiza caesia* L 16.5. Breeds in SE Europe in dry, rocky country with isolated bushes. Resembles Ortolan, but has *blue-grey* (not greenish-grey) *head and breast band,* and rather *orange* (not yellow) *throat*. Female distinguished from male by distinctly streaked crown and faintly streaked breast band. Juvenile is very like that of Ortolan Bunting and in the field they are indistinguishable. Distinguished from juvenile Rock Bunting by more *grey-toned brown* (not reddish-brown) *rump*. Usually seen on the ground. Call a metallic 'spit', sharper than in Ortolan Bunting. Song is similar in structure to a defective or primitive Ortolan song, is thin and lacks the pleasantly ringing tone, invariably has only *one* final note, not two or three. The final note is straight and drawn-out, not ringing or with a diphthong. Two song types are used alternately, one a little stronger and deeper, 'jee jee jee jü', the other higher and almost wheezing, 'weez-weez-weez-wüh'. **V**

Cretzschmar's Bunting

Rock Bunting

♀

♂

Ortolan Bunting

autumn ♀

♂

juv.

Cretzschmar's Bunting

♀

♂

juv.

Yellowhammer

Pine Bunting

Cinereous Bunting

Reed Bunting

Yellowhammer *Emberiza citrinella* L 16.5. Breeds commonly in open country with some bushes, also young conifers, in arable farmland, in juniper country, in clearings in woodland etc. In winter in flocks, feeds in stubble fields, rests in blackthorn scrub, clumps of conifers and the like. Rather long-tailed. In all plumages *rufous-brown rump* and *white on outer tail-feathers* together with some yellow in the plumage, even though some juveniles have little of this. Adult male in spring and summer has *unmarked bright yellow crown patch and throat.* (In winter much of the yellow is concealed by grey-brown and green feather edges.) Female and juvenile less strongly coloured and more streaked. Distinguished from female and juvenile Cirl Bunting by reddish-brown, not grey-toned brown, rump. Call an impure 'steuf' and quiet clicking 'stee**lit**' or 'pitti**lit**' in flight. Song characteristic and well known, varies individually but most often runs 'see-see-see-see-see-see-suuuu'. Sometimes the penultimate note higher than the others ('little-bit-of-bread-and-**no**-cheese'). Some individuals have a more buzzing, River Warbler-like tone, 'dze-dze-dze-dze-...'. RW

Pine Bunting *Emberiza leucocephalos* L 16.5. Very rare vagrant in W Europe from Siberia (breeds from European side of Urals east to Sea of Okhotsk). Has roughly same habitat requirements as Yellowhammer, of which may be said to be eastern counterpart. (In W Siberia, where both species breed side by side within a vast area, a certain amount of inter-breeding occurs; some taxonomists consider them to be races of the same species.) Pronounced migrant, a few wintering regularly in S Europe (e.g. Italy). Male easily recognised by *striking white and red-brown head markings.* Female and juvenile like Yellowhammer (rust-coloured rump), but have *white ground colour to underparts* instead of more or less yellow. Some females have, in addition, a little white on the crown. Calls and song like Yellowhammer's ('little bit of bread' etc). V

Cinereous Bunting *Emberiza cineracea* L 16.5. Very rare and local in Europe, breeding only on Lesbos in Aegean Sea. Visits the island from Mar to Aug. Prefers barren and rocky country. Roughly of Yellowhammer-size, but if anything even more *long-tailed.* Plumage brownish-grey, male with *unmarked, pale greyish-yellow head; breast and flanks grey,* belly white. Female has yellow tinge only on throat, while the head is otherwise grey-brown and diffusely streaked. Juvenile lacks all yellow, is browner and more clearly streaked. Call is both a sharp, scratchy 'tschrip' and a more Ortolan-like 'chü', these often uttered alternately. Song is a simple verse, rather like a subdued Ortolan song, a fairly fast series of ringing notes with some unevenness in the rhythm, e.g. 'zre, zrü-zrü-zrü **zrih**zra'. Typically ends with a high and a low note in rapid succession.

Reed Bunting *Emberiza schoeniclus* L 15.5. Widespread and common species, inhabiting swampy ground with lush lower vegetation, e.g. reedbeds, bushy bogs, etc. Also in drier areas such as conifer planta-tions. In winter also visits cultivated fields, sometimes in company of other buntings and finches, often in small flocks. Male in summer plumage easy to recognise by *white collar, black hood and bib.* White collar is actually mainly a nape band which ends in a wedge up to the bill (white submoustachial stripes). Female has boldly marked head, and pale throat is bordered by *dark lateral throat-stripes.* White outer tail-feathers are visible when bird jerks tail. Can be confused with female and juvenile Rustic Bunting, but told by streaks over breast and on flanks being black, not chestnut-coloured, less prominent wingbars, and dark legs and feet. Slightly bigger than Little Bunting (though this not always easy to see) and also lacks evenly rusty-brown cheeks (see Little Bunting for more details). Call a finely drawn-in, downwards-inflected 'tseeu' and a hoarse 'bzü'. Song rather monotonous, slow and jumpy but varies greatly; a common phrase goes 'tsee tsee, tseea, tsisisirrr'. RWP

juv.

♂

♀

Yellowhammer

♀

♂

Pine Bunting

♂

Cinereous Bunting

♀

Reed Bunting

♀ *winter*

♂ *autumn*

♂ *summer*

Yellow-breasted Bunting

Yellow-breasted Bunting *Emberiza aureola* L 15.5. Common in most of temperate N Asia, also in Russia. Outposts in Finland are now abandoned. Breeds on swampy meadows with osier scrub and scattered trees. Also on bogs in open taiga. Male easily recognised by *black 'face', chestnut-brown crown and breast band* contrasting with *bright yellow underparts* and also by distinct *white wingbars* (almost as Chaffinch), which are also visible in otherwise more subdued winter plumage. First-summer male has smaller white upper wingbar (often streaked dark) and less neat head pattern. Female rather resembles female Yellowhammer, but has duller brown (not rufous-brown), streaked rump, *prominent pale supercilium* and a faint suggestion of paler crown-stripe; median wing-coverts tipped white, forming distinct wingbar; *underparts almost uniformly buffish yellow-white*, streaked only on flanks and sometimes narrowly across breast. Juvenile like female, but has less obvious markings. Call a short clicking 'tick'. Song vaguely recalls Ortolan Bunting's, with notes repeated in twos or threes, 'tru-tru tra-tra **tree-tree**-tra'; typically rises in pitch towards the end, sometimes (like in example) with last note falling. **V**

Cirl Bunting

Cirl Bunting *Emberiza cirlus*. L 16. Breeds in W and S Europe in open country with bushes, hedges and scattered trees, in orchards and near vineyards. Now rare in Britain. In winter seen in open fields, often in flocks together with other buntings and finches. Male is recognised by its *yellow and black head pattern* and *yellow underparts* with *olive-green and rufous breast*. Female is distinguished from very similar female Yellowhammer by *brown-grey, not rufous-brown, rump*, from female Yellow-breasted Bunting by *practically unstreaked rump*, less prominent supercilium and more streaked underparts, from female Ortolan Bunting by more distinct head markings and *grey*, not pink, *bill*. Call a Song Thrush-like 'zit'. Also gives thin, down-slurred 'seeu', and a very fast series of clicking notes, 'zir'r'r'r'r' (like crackling electricity!). Song does not resemble that of other buntings but more of Arctic Warbler song, a rapid, slightly harsh rolling trill, 'zezezezeze…'. Often given from high song post in top of tree. **R**

Black-headed Bunting

Black-headed Bunting *Emberiza melanocephala* L 17. Breeds in SE Europe in open country with scattered bushes and trees, including in cultivated areas. A large bunting with long tail. Male easily recognised by *black head, yellow underparts, chestnut back* and *lack of white tail markings*. Female is rather nondescript, grey-brown and diffusely streaked above, *unmarked pale dirty-yellowish below*; often a *hint of darker grey-brown hood* (corresponding to male's pattern). Note *rather heavy grey bill, no white in tail and pale yellow undertail-coverts*. Juvenile resembles female but has less yellow tinge below. Both female and juvenile very similar to closely related Red-headed Bunting (not treated) of Central Asia; female on average more chestnut-tinged on back (Red-headed tinged yellowish-green, if anything), whereas juveniles cannot be separated in the field. Call a Yellowhammer-like 'cheu' or 'styu', and an Ortolan-like clicking 'plüt'. Song is a short accelerating verse, constantly and indefatigably repeated, rather low-pitched with somewhat jerky rhythm and lots of rolling 'r'-sound in it, 'zrit… zrit… zrüt, zru-zru zütt**ree**-züt zütt**erreh**'. **V**

Yellow-breasted Bunting

Cirl Bunting

Black-headed Bunting

Lapland Bunting

Lapland Bunting (Lapland Longspur) *Calcarius lapponicus* L 15.5. Breeds fairly commonly in open tundra landscape, preferably next to low willow in damp recesses. Also on cloudberry bogs in upper birch zone. A few breed in Scotland on hummocky moors. On migration and in winter seen on stubble fields and dry coastal meadows, mostly singly or in small groups. Spends most of time on ground, where it runs freely. Male in summer plumage unmistakable with black face, throat and breast, *pale yellow streak from the eye backwards*, white neck markings and *reddish-brown nape*. Female more diffusely marked but likewise has fox-red nape, also *pale yellow crown-stripe*. In autumn/winter, sexes and ages similar: head largely rusty-buff, central crown paler brown, underparts whitish with some marks on breast, upperparts boldly streaked, *wings with rufous panel* formed by new greater coverts. *Rusty-red nape* and various amount of black on breast just visible in adults. More robust build and shorter-tailed than most buntings. Call on migration a hard, dry 'prrrt', sometimes followed by a short 'chu' (may at times be very like Snow Bunting's 'pew'), alternatively harsh 'jeeb' (also heard from night migrants). At breeding site metallic 'teehü' is commonest call (alarm?). Song short and jingling, similar to Horned Lark (when delivered in short phrases); also like Snow Bunting, but jingling elements characteristic, and phrases are repeated fairly constantly and are not so varied; e.g. 'kretle-**krlee**-trr kritle-kretle-trü'. Has song flight, in which descends on outspread wings. **WP**

Snow Bunting

Snow Bunting *Plectrophenax nivalis* L 16. Breeds fairly commonly at high altitude in mountains among rocks and patches of snow but also at sea level in the Arctic. In winter, visits open coastal areas or extensive arable plains. Male in summer *white and black*; in particular, wings (secondaries and coverts) *shining white*. Female has dirty grey-brown head, grey-black back and wing-coverts, white secondaries. In autumn and winter both sexes characterised by broad *buff feather fringes on upperparts* and rusty on head, cheeks and breast, but in flight still very white. Juveniles (Jul–Aug) have grey head and brown-spotted breast. By autumn they have moulted into a plumage resembling that of the adults. On average, though, they have less white in the wing; the immature female has only bases of secondaries white, forming a broad, white, translucent wing bar. Flies in long undulations, the flocks glistening white. Call a short but full whistle, 'pew' (usually from single birds) and a softly twittering 'dirrirr**it**', recalling Crested Tit, softer than similar call of Lapland Bunting. Sand Martin-like rasping calls are heard from large flocks. The song is short and clear, with resemblances to both Rustic and Lapland Buntings (though lacks latter's jingling tone), e.g. 'sweeto-süway-weetüta-süwee'. It is given from boulder or in descending gliding flight. **RWP**

Sparrows (family Passeridae)

Sparrows are described on p.276 apart from the Snow Finch (in spite of the name a sparrow!), which is dealt with here for easier comparison with Snow Bunting.

Snow Finch

Snow Finch *Montifringilla nivalis* L 18. Breeds in the Alps and S Europe on high mountains above the tree line and below the snow. Nests in crevices and under boulders. Outside breeding season often seen in small flocks. In winter moves to lower level, but can certainly be seen in Feb at the top stations of ski resorts. In all plumages easily distinguished from Snow Bunting by *grey head* (exception: juvenile Snow Bunting in Jul–Aug has grey head). Male in summer has a *small blackish bib*, female some grey visible on chin. Note *white markings on wings and tail*, less extensive in female and juvenile. Juvenile resembles female. Bill yellow in winter, black (male) or dark with paler base (female) in summer. Often jerks tail. Calls include a somewhat hoarse 'tseeh' and a purring 'prrt'. The song, uttered from perch on a boulder, or in gliding song-flight, is very jerky and jolting, somewhat recalling Twite's.

Lapland Bunting

♀ summer

♂ summer

♂ winter

Snow Bunting

♂ winter

♀

♂ summer

♀

♀ winter

♂ summer

♂

Snow Finch

Corn

Little

Ortolan

juv.

Cretzchmar's

juv.

Reed

♀

juv.

Yellowhammer

Cirl

♀

♀

Black headed

♀
Rustic

juv.
Rock

♀ *winter*
Lapland

♀
Pine

♀
Cinereous

♀
Yellow-breasted

♀ *juv.*

♀ *winter*
Snow

BIBLIOGRAPHY

The following is a selected list of references, mainly of modern books on bird identification covering Europe, and of regional atlases and checklists. A few older books, or books covering other areas, have also been included because of their importance.

Alström, P., Colston, P. & Lewington, I. (1991) *A Field Guide to the Rare Birds of Britain and Europe*. Domino, Jersey.

Alström, P. & Mild, K. (2003) *Pipits & Wagtails of Europe, Asia and North America*. Helm, London.

Beaman, M. (1994) *Palearctic Birds*. Stonyhurst.

Beaman, M. & Madge, S. (1998) *The Handbook of Bird Identification for Europe and the Western Palearctic*. Helm, London.

Bergmann, H.-H. & Helb, H.-W. (1982) *Stimmen der Vögel Europas*. BLV, Munich.

BWP Update. (1997–2004) *The Journal of Birds of the Western Palearctic*. Oxford University Press, Oxford.

Campbell, B. & Lack, E. (eds.) (1985) *A Dictionary of Birds*. Poyser, Calton.

Chantler, P. & Driessens, G. (1995) *Swifts*. Pica, Mountfield.

Cleere, N. & Nurney, D. (1998) *Nightjars*. Pica, Mountfield.

Clement, J. F. (2000) *Birds of the World*. A Checklist. 5th ed. Pica, Mountfield.

Clement, P., Harris, A. & Davis, J. (1993) *Finches and Sparrows*. Helm, London.

Clement, P. & Hathway, R. (2000) *Thrushes*. Helm, London.

Constantine, M. et al. (2006) *The Sound Approach to birding*. With 2 CD. Poole.

Cottridge, D. & Vinicombe, K. (1996) *Rare Birds in Britain & Ireland*. A photographic record. HarperCollins, London.

Cramp, S., Simmons, K. E. L. & Perrins, C. M. (eds.) (1977–94) *Handbook of the Birds of Europe, the Middle East and North Africa* ('The Birds of the Western Palearctic'). 9 vols. Oxford University Press, London.

Cramp, S. et al. (eds.) (2006) *BWPi*, version 2.0. (Electronic version of the 9-vol. handbook, the Concise edition and the BWP Updates, plus added film footage and sounds.) Birdguides, London.

Curry-Lindahl, K. (ed.) et al. (1959–62) *Våra fåglar i Norden*. ('Our Nordic Birds.') 4 vols. 2nd ed. Natur och Kultur, Stockholm.

Delacour, J. & Scott, P. (1954–64) *The Waterfowl of the World*. 4 vols. Country Life, London.

Delin, H. & Svensson, L. (1988) *Photographic Guide to the Birds of Britain & Europe*. Hamlyn, London.

Dement'ev, G. P. & Gladkov, N. A. (eds.) et al. (1951–54) *Birds of the Soviet Union*. 6 vols. (Israel program for scientific translations, Jerusalem 1966–70.)

Dickinson, E. C. (ed.) et al. (2003) *The Howard & Moore Complete Checklist of the Birds of the World*. 3rd ed. Helm, London.

Dubois, P. J., Le Marechal, P., Olioso, G., Yésou, P. et al. (2000) *Inventaire des Oiseaux de France*. Nathan, Paris.

Dunn, J. L. (ed.) et al. (1983, 1999) *Field Guide to the Birds of North America*. 3rd ed. National Geographic Society, Washington.

Ferguson-Lees, J. & Christie, D. A. (2001) *Raptors of the World*. Helm, London.

Ferguson-Lees, J., Willis, I. & Sharrock, J. T. R. (1983) *The Shell Guide to the Birds of Britain and Ireland*. Michael Joseph, London.

Fjeldså, J. (1977) *Guide to the Young of European Precocial Birds*. Skarv, Tilsvide.

Flint, V. E., Boehme, R. L., Kostin, Y. V. & Kutznetsov, A. A. (1984) *A Field Guide to Birds of the USSR*. Princeton University Press, Princeton.

Forsman, D. (1999) *The Raptors of Europe and the Middle East*. Poyser, London.

Fry, C. H., Fry, K. & Harris, A. (1992) *Kingfishers, Bee-eaters & Rollers*. Helm, London.

Génsbøl, B. (1984) *Birds of Prey of Europe*. Harper Collins, London.

Géroudet, P. (1953–61) *Les Passereaux d'Europe*. 3 vols. Delachaux et Niestlé, Neuchâtel and Paris.

Géroudet, P. (1965) *Les Rapaces Diurnes et Nocturnes d'Europe*. Délachaux et Niestlé, Neuchâtel and Paris.

Géroudet, P. (1965) *Water-birds with webbed feet*. Blandford, London.

Gibbs, D., Barnes, E. & Cox, J. (2001) *Pigeons and Doves*. Pica, Mountfield.

Gill, F. & Wright, M. (2006) *Birds of the World*. Recommended English names. IOC and Helm, London.

Gjershaug, J. O., Thingstad, P. G., Eldøy, S. &

Byrkjeland, S. (ed.) (1994) *Norsk fugleatlas.* NOF, Klæbu.

Glutz von Blotzheim, U. N. (publ.), Bauer, K. & Bezzel, E. (1966–98) *Handbuch der Vögel Mitteleuropas.* 14 vols. plus Index vol. Aula, Wiesbaden.

Gooders, J. (1986) *Field Guide to the Birds of Britain & Ireland.* Kingfisher Books, London.

Gorman, G. (1996) *The Birds of Hungary.* Helm, London.

Grant, P. J. (1986) *Gulls: a guide to identification.* 2nd ed. Poyser, Calton.

Haftorn, S. (1971) *Norges fugler.* Universitets Forlaget, Trondheim.

Hagemeijer, W. J. M. & Blair, M. J. (ed.) (1997) *The EBCC Atlas of European Breeding Birds.* Poyser, London.

Hancock, J. & Kushlan, J. (1984) *The Herons Handbook.* Harper & Row, London and New York.

Handrinos, G. & Akriotis, T. (1997) *The Birds of Greece.* Helm, London.

Harrap, S. & Quinn, D. (1996) *Tits, Nuthatches & Treecreepers.* Helm, London.

Harris, T. & Franklin, K. (2000) *Shrikes & Bush-Shrikes.* Helm, London.

Harrison, P. (1989) *Seabirds of the World.* Helm, London.

Harrison, C. (1982) *An Atlas of the Birds of the Western Palearctic.* Collins, London.

Harrison, P. (1986) *Seabirds: an identification guide.* 2nd ed. Helm, Beckenham.

Hartert, E. (1910–22) *Die Vögel der paläarktischen Fauna.* (Suppl. vols. 1923, 1932–36.) Friedländer und Sohn, Berlin.

Hayman, P., Marchant, J. & Prater, T. (1986) *Shorebirds.* Helm, London.

Hayman, P., Hume, R. (2001) *The Complete Guide to the Birdlife of Britain and Europe.* Mitchell Beazley, London.

Heinzel, H., Fitter, R. & Parslow, J. (1995) *Birds of Britain & Europe with North Africa & the Middle East.* 2nd ed. Harper Collins, London.

Hollom, P. A. D., Porter, R. F., Christensen, S. & Willis, I. (1988) *Birds of the Middle East and North Africa.* Poyser, London.

del Hoyo, J., Elliott, A. & Sargatal, J. (eds.) (1992–) *Handbook of the Birds of the World.* Vol. 1–11. Lynx, Barcelona.

Hume, R. (2002) *Birds of Britain and Europe.* Dorling Kindersley, London.

Jenni, L. & Winkler, R. (1994) *Moult and Ageing of European Passerines.* Academic Press, London.

Il'icev, V. D., Flint, V. E. et al. (1985–) *Handbuch der Vögel der Sowjetunion.* Vols. 1, 4, 6:1. (In German. Unfinished handbook project.) NBB, Wittenberg Lutherstadt.

Jonsson, L. (1992) *Birds of Europe with North Africa and the Middle East.* Helm, London.

King, B. F., Dickinson, E. C. & Woodcock, M. W. (1975) *A Field Guide to the Birds of South-East Asia.* Collins, London.

König, C., Weick, F. & Becking, J.-H. (1999) *Owls. A Guide to the Owls of the World.* Pica, Mountfield.

Kren, J. (2000) *Birds of the Czech Republic.* Helm, London.

Laine, L. J. (2004) *Suomalainen Lintuopas.* ('Finnish Birds'.) Rev. ed. WSOY, Helsinki.

Larsson, L., Larsson, E. & Ekström, G. (2002) *Birds of the World.* Interactive electronic checklist, 2 CD. Väse, Sweden.

Lefranc, N. & Worfolk, T. (1997) *Shrikes.* Pica, Mountfield.

Leibak, E., Lilleleht, V. & Veromann, H. (1994) *Birds of Estonia.* Estonian Academy Publ., Tallinn.

Lindell, L. (ed.) et al. (2002) *Sveriges fåglar.* ('The Birds of Sweden'.) 3rd ed. SOF, Stockholm.

Madge, S. & Burn, H. (1988) *Wildfowl.* Helm, London.

Madge, S. & Burn, H. (1991) *Crows and Jays.* Helm, London.

Madge, S. & McGowan, P. (2002) *Pheasants, Partridges & Grouse.* Helm, London.

Martí, R. & del Moral, J. C. (ed.) (2003) *Atlas de las Aves Reproductoras de España.* ('Atlas of Breeding Birds in Spain'.) Ministerio de Medio Ambiente/SEO Birdlife, Madrid.

Mebs, T. & Scherzinger, W. (2000) *Die Eulen Europas.* Kosmos, Stuttgart.

Meschini, E. & Frugis, S. (ed.) (1993) *Atlante degli Uccelli Nidificanti in Italia.* ('Atlas of Breeding Birds in Italy'.) Suppl. Ric. Biol. Selvaggina, Bologna.

Mikkola, H. (1983) *Owls of Europe.* Poyser, Calton.

Mitchell, D. & Young, S. (1997) *Rare Birds of Britain and Europe.* New Holland, London.

Olsen, K. M. & Larsson, H. (1995) *Terns of Europe and North America.* Helm, London.

Olsen, K. M. & Larsson, H. (1997) *Skuas and Jaegers*. Pica, Mountfield.

Olsen, K. M. & Larsson, H. (2003) *Gulls of Europe, Asia and North America*. Helm, London.

Olsson, U., Curson, J. & Byers, C. (1995) *Buntings and Sparrows*. Pica, Mountfield.

Peterson, R. T., Mountfort, G. & Hollom, P. A. D. (1993) *A Field Guide to the Birds of Britain and Europe*. 5th ed. Harpers Collins, London.

Porter, R. F., Christensen, S., Nielsen, B. P. & Willis, I. (1981) *Flight identification of European Raptors*. 3rd ed. Poyser, Calton.

Porter, R. F., Christensen, S. & Schiermacker-Hansen, P. (1996) *Field Guide to the Birds of the Middle East*. Poyser, London.

Priednieks, J., Strazds, M., Strazds, A. & Petrins, A. (1989) *Latvijas Ligzdojoso Putno Atlants 1980–1984*. Zinatne, Riga. (With English summaries.)

Pyle, P. et al. (1997) *Identification Guide to North American Birds*. Part 1. Slate Creek Press, Bolinas, California.

Prater, A. J., Marchant, J. & Vuorinen, J. (1977) *Guide to the Identification and Ageing of Holarctic Waders*. BTO Guide no. 17. BTO, Tring.

Rosenberg, E. (1953, 1972) *Fåglar I Sverige*. ('Birds of Sweden.') 5th ed. Almqvist & Wiksell, Uppsala.

Rufino, R. (ed.) et al. (1989) *Atlas das Aves que nidificam em Portugal Continental*. Minist. plano e administr. território, Lissabon.

Ryabtsev, V. K. (2001) *Ptitsy Urala*. ('The Birds of Ural'.) Jekaterineburg.

Scott, P. (1968) *A Coloured Key to the Wildfowl of the World*. Wildfowl Trust, Slimbridge.

Shirihai, H., Christie, D. & Harris, A. (1996) *Birder's Guide to European and Middle Eastern Birds*. MacMillan, London.

Shirihai, H., Gargallo, G. & Helbig, A. J. (2001) *Sylvia Warblers*. Helm, London.

Sibley, D. A. (2000) *The Sibley Guide to Birds*. Knopf, New York.

Snow, D. & Perrins, C. M. (1998) *The Birds of the Western Palearctic. Concise Edition*. 2 vols. Oxford University Press, Oxford.

Stresemann, E., Portenko, L. A., Dahthe, H., Neufeldt, I. A. et al. (1960–1998) *Atlas der Verbreitung palaearktischer Vögel*. 17 vols. Akademie-Verlag, Berlin.

Svensson, L. (1992) *Identification Guide to European Passerines*. 4th ed. Lullula, Stockholm.

Svensson, L., Grant, P. J., Mullarney, K. & Zetterström, D. (1999) *Collins Bird Guide*. Harper Collins, London.

Svensson, S., Svensson, M. & Tjernberg, M. (1999) *Svensk fågelatlas*. ('Atlas of Swedish Birds'.) Vår Fågelvärld, Suppl. 31. SOF, Stockholm.

Taylor, B. & van Perlo, B. (1998) *Rails*. Pica, Mountfield.

Ticehurst, C. B. (1938) *A Systematic Review of the Genus Phylloscopus*. British Museum (Nat. Hist.), London.

Tomialojc, L. & Stawarczyk, T. (2003) *Awifauna Polski*. (The Avifana of Poland. Distribution, numbers and trends.) 2 vols. PTPP, Wroclaw.

Turner, A. & Rose, C. (1989) *Swallows and Martins of the World*. Helm, London.

Urquhart, E. (2002) *Stonechats*. A Guide to the Genus *Saxicola*. Ill. by A. Bowley. Helm, London.

Vaurie, C. (1959, 1966) *The Birds of the Palearctic Fauna*. 2 vols. Witherby, London.

Vinicombe, K., Harris, A. & Tucker, L. (1989) *Bird Identification*. MacMillan, London.

Winkler, H., Christie, D. A. & Nurney, D. (1995) *Woodpeckers*. Pica, Mountfield.

Voous, K. H. (1977) *List of Recent Holarctic Bird Species*. Ibis, Suppl. London.

Williamson, K. (1966–67) *Identification for Ringers*. 3 vols. BTO, Tring.

Witherby, H. F., Jourdain, F. C. R., Ticehurst, N. & Tucker, B. W. (1938–41) *The Handbook of British Birds*. 5 vols. Witherby, London.

Zink, G. (1973–85) *Der Zug europäischer Singvögel*. 4 vols. Vogelzug-Verlag, Möggingen.

A large number of magazines and journals treating various aspects of bird life and bird-watching activities are now published. Almost every country in Europe has its own selection of such journals of varying ambitions and importance. Some of the most useful are listed below, together with the publishers where these are societies, museums or institutions. Where no publisher is mentioned, the journal is a commercial one.

Austria	Egretta	Österreichischen Vogelwarte
Belgium	Aves	Société d'Études Ornithologiques
	Gerfaut	Société Ornithologique de la Belgique
Britain	Birding World	–
	Birds	Roy. Soc. for the Protection of Birds (RSPB)
	Bird Study	British Trust for Ornithology (BTO)
	Birdwatch	–
	Bird Watching	–
	British Birds	–
	Bulletin B. O. C.	British Ornithologists' Club
	Ibis	British Ornithologists' Union (BOU)
	Ringing & Migration	BTO
	Sandgrouse	Orn. Soc. of the Middle East, Caucasus and Central Asia (OSME)
Denmark	Dansk Ornitologisk Forenings Tidsskrift	Dansk Ornitologisk Forening (DOF)
	Fugle i Felten	DOF
Finland	Alula	–
	Lintumies	Ornitologiska Föreningen I Finland (OFF)
	Linnut	BirdLife Finland
	Ornis Fennica	(OFF)
France	Alauda	Société d'Études Ornithologiques
	L'Oisseau	Société Ornithologique de France
Germany	Journal of Ornithology	Deutsche Ornithologen-Gesellschaft (DOG)
	Limicola	–
	Die Vogelwarte	DOG
	Die Vogelwelt	–
Ireland	Irish Birds	Irish Wildbird Conservancy
Italy	Avosetta	Associazone Ornitologica Italiana
	Rivista Italiana di Ornithologia	–
Netherlands	Ardea	Nederlandse Ornithologische Unie (NOU)
	Dutch Birding	–
	Limosa	NOU
Norway	Vår Fuglefauna	Norsk Ornitologisk Forening (NOF)
	Ornis Norvegica	NOF
Poland	Acta Ornithologica	Musei Zoologici Polonici
Spain	Ardeola	Sociedad Española de Ornitologia
	La Garcilla	SEO/BirdLife International
Sweden	Vår Fågelvärld	Sveriges Ornitologiska Förening (SOF)
	Ornis Suecica	SOF
	Roadrunner	Club 300
Switzerland	Nos Oiseaux	Soc. romande pour l'étude et la prot. des oiseaux
	Der Ornithologische Beobachter	Schweizerische Gesellsch. für Vogelkunde und Vogelschutz

BIRD VOICE RECORDINGS

Recording bird song and calls has for a long time being a rather specialised activity among amateur birders. With the arrival of simpler and cheaper tools it is likely to gain in popularity in years to come, not least since it gives important help in identification of many difficult species. But regardless whether you plan to obtain your own recordings or not, you will probably want to acquire a few selections of recordings available on the market. The following is meant as guide.

Andersson, B., Svensson, L. & Zetterström, D. (1990) *Fågelsång i Sverige*. 1 CD and book (in Swedish). (90 common Swedish birds.) Mono Music, Stockholm.

Chappuis, C. (1987) *Migrateurs et hivernants*. 2 cassettes. Grand Couronne, France.

Chappuis, C. (2000) *African Bird Sounds – 1*. North-West Africa, Canaria and Cap-Verde Islands. 4 CD. SEOF/Nat. Sound Arch., London.

Constantine, M. *et al.* (2006) *The Sound Approach to birding*. With 2 CD. Poole.

Cramp, S. *et al.* (eds.) (2006) *BWPi*, version 2.0. (Electronic version of the 9-vol. handbook, the Concise edition and the BWP Updates, plus added film footage and sounds.) Birdguides, London.

Gulledge, J. (prod.) *et al.* (1983) *A Field Guide to Bird Songs of Eastern and Central North America*. 2nd ed. 2 cassettes. Cornell/Houghton Mifflin, Boston.

Gunn, W. W. H., Kellogg, P. P. (ed.) *et al.* (1975) *A Field Guide to Western Bird Songs*. 3 cassettes. Cornell/Houghton Mifflin, Boston.

Jännes, H. (2003) *Calls of Eastern Vagrants*. 1 CD. Earlybird, Helsinki.

Mild, K. (1987) *Soviet Bird Songs*. 2 cassettes and booklet. Stockholm.

Mild, K. (1990) *Bird Songs of Israel and the Middle East*. 2 cassettes and booklet. Stockholm.

Palmér, S. & Boswall, J. (1981) *A Field Guide to the Bird Songs of Britain and Europe*. 16 cassettes. SR Phonogram, Stockholm.

Ranft, R. & Cleere, N. (1998) *A Sound Guide to Nightjars and related Nightbirds*. 1 CD. Pica/Nat. Sound Arch., London.

Robb, M. S. (2000) *Introduction to vocalizations of crossbills in north-western Europe*. 1 CD and leaflet. Dutch Birding, Amsterdam.

Roché, J. C. (1998) *Birds of prey and Owls of Western Europe*. 1 CD. Frémeaux, Cervennes.

Roché, J. C. *et al.* (1990) *All the bird songs of Britain and Europe*. 4 CD. Sittelle, Mens.

Roché, J. C. & Chevereau, J. (2002) *Bird Sounds of Europe & North-west Africa*. 10 CD. Wildsounds, Salthouse.

Sample, G. (1998) *Bird Call Identification*. Book and 1 CD. HarperCollins, London.

Sample, G. (2003) *Warbler Songs & Calls of Britain and Europe*. Book and 3 CD. HarperCollins, London.

Schubert, M. (1981) *Stimmen der Vögel Zentralasiens*. 2 LP records. Eterna.

Schubert, M. (1984) *Stimmen der Vögel*. VII. Vogelstimmen Südosteuropas (2). 1 LP record. Eterna.

Schulze, A. (ed.), Roché, J. C., Chappuis, C., Mild, K. et al. (2003) *Die Vogelstimmen Europas, Nordafrikas und Vorderasiens*. 17 CD. Edition Ample, Germering.

Songs and calls of British and European birds. (1990) Kettle, R. & Svensson, L. (eds.) 3 cassettes. Nat. Sound Arch. and Hamlyn, London.

Strömberg, M. (1994) *Moroccan Bird Songs and Calls*. 1 cassette. Sweden.

Svensson, L. (1984) *Soviet Birds*. 1 cassette. Stockholm.

Veprintsev, B. N. *et al.* (1982–86) *Birds of the Soviet Union: A Sound Guide*. 7 LP records. Melodia, Moscow.

Wahlström, S. (1995) *Från Alfågel till Ärtsångare*. 2 CD. Wiken.

Walton, R. K. & Lawson, R. W. (1994) *More Birding by Ear*. A Guide to Bird-song Identification. Eastern/Central North America. 3 CD and booklet. Houghton Mifflin, Boston.

Wetland Birds – a celebration. (1996) 1 CD. Wildfowl & Wetlands Trust/Bird Watching, England.

ARTISTS' ACKNOWLEDGEMENTS

Håkan Delin: pages 8–17; 27; 29; 33 (Green-winged Teal, Garganeys, flying top two); 41 (Goldeneye and Barrow's Goldeneye, heads); 51 (male Capercaillie and Hazel Grouse); 57; 73 (egrets, flying); 75 (Grey and Purple Herons, flying); 79; 99 (Gyr Falcons); 107; 109; 167; 169; 173; 175 (Cuckoo, flying, song posture, young with foster parent); 177; 179; 181; 189; 191; 193; 201 (Horned Lark, juv.); 205 (Olive-backed Pipit); 211 (Grey Wagtail); 225 (Black-throated Thrushes); 231 (Lanceolated Warbler); 251 (Dusky and Radde's Warblers); 253 (Yellow-browed and Pallas's Warblers); 255 (Semi-Collared Flycatcher); 263 (Krüper's Nuthatch); 267 (Southern Grey Shrike, flying figures except flying Red-backed Shrike); 298 (Little Bunting); 299 (Lapland Bunting and Snow Bunting, juv.)

Martin Elliott: pages 37; 39; 59; 61; 63; 65; 149 (Heuglin's Gulls); 151

Peter Hayman: pages 185; 197; 199; 201 (all except Horned Lark, juv.); 203; 205 (except Olive-backed Pipit); 207; 213; 215 (Robin); 219 (Isabelline Wheatear, flying); 225 (Blackbird and Ring Ouzel); 231 (Savi's, River and Grasshopper Warblers); 233; 235; 237; 239; 241; 243; 245; 247 (Rufous Bush Robin); 249; 251 (Western Bonelli's, Eastern Bonelli's, Arctic and Greenish Warblers); 253 (Goldcrest and Firecrest); 255 (Collared and Semi-collared Flycatchers, flying) 257; 259; 273 (flying figures); 275 (flying figures); 287; 289 (Little and Rustic Buntings; Corn Bunting, flying); 293 (Reed Bunting, autumn and winter)

Arthur Singer: pages 20–21; 23; 25; 31; 33 (except Green-winged Teal and Garganeys, flying top two); 35; 41 (except Goldeneye and Barrow's Goldeneye heads); 43; 45; 47; 49; 51 (except male Capercaillie and Hazel Grouse); 53; 55; 67; 69; 71; 73 (except egrets, flying); 75 (except Grey and Purple Herons, flying); 77; 80; 83; 85; 87; 89; 91; 93; 95; 97; 99 (except Gyr Falcons); 101; 103; 104-105; 111; 113; 171; 175 (except Cuckoo, flying, song posture, young with foster parent); 183; 187; 194-195; 209; 211 (except Grey Wagtail); 215 (except Robin); 217; 219 (except Isabelline Wheatear, flying); 221; 223; 227; 228-229; 247 (except Rufous Bush Robin); 255 (Pied, Collared, Red-Breasted and Spotted Flycatchers); 261; 263 (except Krüper's Nuthatch); 265; 267 (except Southern Grey Shrike and flying figures except flying Red-backed Shrike); 269; 271; 273 (except flying figures); 275 (except flying figures); 277; 279; 281; 283; 285; 289 (Corn Bunting, perched); 291; 293 (except Reed Bunting, autumn and winter); 295; 297; 298 (except Little Bunting); 299 (except Lapland Bunting and Snow Bunting, juv.)

Lars Svensson: All black and white illustrations; raptors in flight pages 83, 87, 89, 91, 97

Dan Zetterström: pages 115, 117, 119, 121, 123, 125, 127, 129, 131, 133, 135, 137, 139, 141, 143, 145, 147, 149 (except Heuglin's Gull); 153, 155; 157; 161; 163; 165

SCIENTIFIC NAMES INDEX

The index below lists species by their binomial scientific (Latin) name. An index of common names can be found starting on page 312. A **Quick Index** for use in the field can be found on page 320.

COMMON NAME INDEX

The index below lists species by their common name. An index of scientific binomial names can be found starting on page 306. A **Quick Index** for use in the field can be found on page 320. The page reference is to the species description, the illustration appears on the opposite page.

QUICK INDEX

This index is intended to help rapid identification when in the field. For more complete lists of species, both common and scientific, please refer to the indexes on the previous pages (306–319).